W0042346

Scientific Computing on Supercomputers

Scientific Computing on Supercomputers

Edited by

Jozef T. Devreese and
Piet E. Van Camp

University of Antwerp
Antwerp, Belgium

PLENUM PRESS • NEW YORK AND LONDON

Library of Congress Cataloging in Publication Data

International Workshop on the Use of Supercomputers in Theoretical Science
Scientific computing on supercomputers / edited by Jozef T. Devreese and Piet E. Van Camp.
p. cm.
"Proceedings of the Second, Third, and Fourth International Workshops on the Use of Supercomputers in Theoretical Science, held December 12, 1985, June 16, 1987, and June 9, 1988, at the University of Antwerp, Antwerp, Belgium"—T.p. verso.
Includes bibliographies and index.

ISBN-13: 978-1-4612-8098-9 e-ISBN-13: 978-1-4613-0819-5
DOI: 10.1007/978-1-4613-0819-5

1. Supercomputers—Congresses. 2. Science—Data processing—Congresses. I Devreese, J. T. (Jozef T) II. Van Camp, P. E. (Piet E.) III. Title.
QA76.5.I623 1989 89-8428
004.1'1—dc20 CIP

Proceedings of the Second, Third, and Fourth International Workshops on the Use of Supercomputers in Theoretical Science, held December 12, 1985, June 16, 1987, and June 9, 1988, at the University of Antwerp, Antwerp, Belgium

Softcover reprint of the hardcover 1st edition 1989

A Division of Plenum Publishing Corporation
233 Spring Street, New York, N.Y. 10013

PREFACE

The International Workshops on "The Use of Supercomputers in Theoretical Science" have become a tradition at the University of Antwerp, Belgium. The first one took place in 1984.

This volume combines the proceedings of the second workshop (December 12, 1985), of the third (June 16, 1987) and of the fourth (June 9, 1988).

The principal aim of the International Workshops is to present the state-of-the-art in scientific high speed computation. Indeed, during the past ten years computational science has become a third methodology with merits equal to the theoretical and experimental sciences. Regretfully, access to supercomputers remains limited for academic researchers. Nonetheless, supercomputers have become a major tool for scientists in a wide variety of scientific fields, and they lead to a realistic solution of problems that could not be solved a decade ago.

It is a pleasure to thank the Belgian National Science Foundation (NFWO-FNRS) for the sponsoring of all the workshops. These workshops are organized in the framework of the Third Cycle "Vectorization, Parallel Processing and Supercomputers", which is also funded by the NFWO-FNRS. The other sponsor I want to thank is the University of Antwerp, where the workshops took place. The University of Antwerp (UIA), together with the NFWO-FNRS, are also the main sponsors of the ALPHA-project, which gives the scientists of Belgium the opportunity to obtain an easy supercomputer connection.

Special thanks are due to Mrs. H. Evans for the skillful typing of the manuscripts and for the careful way in which she prepared the author and subject index.

J.T. Devreese
Professor of Theoretical Physics

December 1988

CONTENTS

I. SUPERCOMPUTER ARCHITECTURES

II. SUPERCOMPUTER LANGUAGES AND ALGORITHMS

III. SUPERCOMPUTER APPLICATIONS

IV. INDEXES

I. SUPERCOMUTERS ARCHITECTURES

PERFORMANCE MODELLING OF SUPERCOMPUTER ARCHITECTURES AND ALGORITHMS

Erik H. D'Hollander

Department of Electrical Engineering, State University of Ghent, St.-Pietersnieuwstraat 41, B-9000 GHENT, Belgium

ABSTRACT

The need for more processing power has spurred the development of new processor architectures. While these machines bear a high speed potential, this speedup over a conventional machine is only partly achieved, depending on the type of calculations. In this paper an overview is given of the blend of techniques which allow to estimate the processing rate for various existing computer architectures in function of their serial versus vector calculations.

I. INTRODUCTION: THE SPECTRUM OF COMPUTERS

In the last decade the computer world has grown at an exponential rate giving rise to a vast selection of computer-architectures, going from personal computers to supercomputers.

The most spectacular advance is noted in the <u>microprocessor</u> and personal computer market, where the results of <u>microelectronics</u> for a long time have led to immediately successful products, although this market now tends to become saturated. The speed at which this micro-(r)evolution took place is not only due to economical factors, such as the decreasing price of integrated circuits, but also to the <u>existing know-how</u> to build larger computers. This made it possible to greatly reduce the development of these new architectures, both hardware and software-wise and has caused the proliferation of so many microprocessor companies.

At the other end of the spectrum the impact of <u>supercomputers</u> on the computer world is not so spectacular. Although supercomputing companies advertise gigantic computing speeds with respect to the conventional machines, their products are not so well accepted and recently one major company had to close doors. This lack of a breakthrough can be attributed to several factors.

First parallel machines are not userfriendly. Although the high performance can be achieved in very particular applications, it requires much effort of the programmer to restructure his program in order to take advantage of the fastest components in the hardware. This causes many potential users to turn to conventional machines and wait a little bit more for their results.

Second, the price of most supercomputers is prohibitive and limits the potential buyers to large companies or scientific institutes. Even the poor-man's Floating Point Systems array processor costs several hundred thousands of dollars. Faster systems such as the CRAY or CDC CYBER computers additionally require a suitable environment and a power supply of up to 250 KW, largely used to produce and remove heat energy.

Third, supercomputers are not general purpose machines. Most of them require a front-end computer in order to provide a user-interface and to support peripheral equipment. In fact the supercomputer is merely used for computationally expensive problems, where manifestly the weakest resource is the processor. Some machines such as the CYBER 205 have extra hardware for string-manipulation, thereby opening a gate towards database applications, but in general the hardware or software of existing supercomputers is not oriented towards general purpose tasks and therefore differs conceptually from Japan's fifth generation project, where symbolic processing, artificial intelligence and database techniques are incorporated in a multifunctional design.

Nevertheless, supercomputers emerge and apparently fill an existing need, i.e. the need for more processing power.

II. THE NEED FOR FAST PROCESSING AND ITS SOLUTIONS

It seems to be an unwritten law that the available processing time of any computer tends to zero as its lifetime increases. In other words, when the processing power is there, more complex problems are solved and their solution generates new scientific and technological problems, which in their turn require more computer time [1]. This lemma is applicable to most computer environments, from CAD to Artificial Intelligence, and also forms a solid base for the development of modern supercomputing systems.

2.1. Increasing the Memory Speed

The communications path between the processor and the memory for fetching instructions and reading or writing data limits the amount of work that can be done between two memory accesses. An instruction normally involves the following phases:

IF = Instruction Fetch	Memory action
ID = Instruction Decode	Processor action
OF = Operand(s) Fetch	Memory action
EX = Execute	Processor action

Clearly, the speed of execution depends to a large extent on the memory bandwidth. A first way to enhance the processing

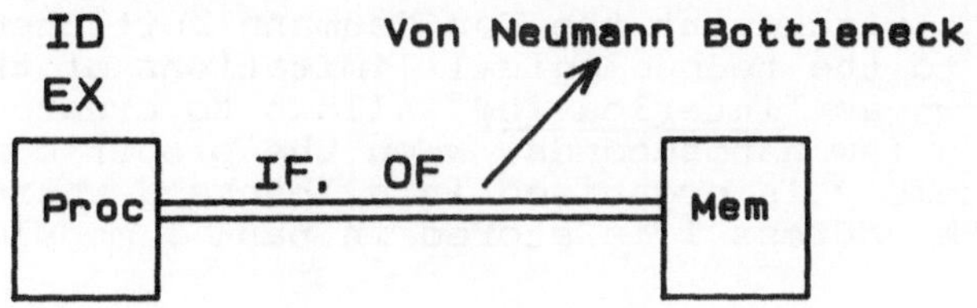

Fig. 1. Stored Program Concept.

power is to increase the memory accessibility through a <u>wider databus</u>. Whereas in conventional computers the databus is 16 to 32 bits wide, supercomputers use a 64 (FPS), 128 (CRAY) or even 512 bit wide datapath (CDC Cyber 205).

In conjunction with a wider bus, <u>faster memory</u> is used. Since most supercomputers obtain their maximal speed on processing large arrays of data, memory speed is of prime importance. Currently minimal memory access times are of the order of 50 ns.

However, in many architectures a datarate of a few nanoseconds per memory fetch is required. Consequently, with today's technology and memory sizes, there is no economical way for addressing a Mega- or Gigabyte memory at the required speed.

The slow processor-memory interconnection has been commonly called the Von Neumann bottleneck, after the inventor of the Stored Program concept (Fig. 1).

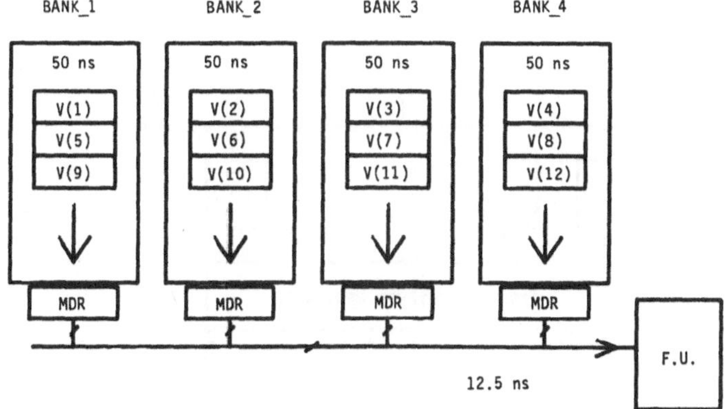

Fig. 2. Interleaved Memory Access with 4 Memory Banks.

In order to circumvent the Von Neumann bottleneck, which is largely due to the technological limitations of the memory, a technique known as "interleaving" allows to create a 'virtual' access time of a few nanoseconds, when the proper conditions are met. The memory is organized in n "banks", where the data at location with address i is stored in bank i modulo(n) (Fig. 2).

This organisation allows to access consecutive array elements within 20 ns (CDC Cyber 205), or even 12.5 ns (CRAY-1), provided the data are stored at consecutive addresses.

2.2. Increasing the Processor Speed

In a balanced system, the processor speed has to match the speed of the memory. Increasing the processor speed starts by using a faster technology e.g. by using ECL, higher integration, greater clock rate, smaller dimensions in order to obtain a lower propagation time, and the like. All these measures have been taken in advanced machines, but they are very costly and failed in the end to reach the megaflop (million of floating point operations per second).

The only way to increase the processing power beyond the physical and economical limits of conventional architectures, is to operate different processors in parallel on the same program. When we exclude the experimental dataflow architectures, i.e. restrict ourselves to a program-driven instead of a data-driven execution, then the well-known Flynn topology [2] is helpful in classifying the existing supercomputer architectures. Flynn discerns single (S) or multiple (M) instruction (I) and data (D) streams, and forms four theoretical architectures, of which the closest real architectures are:

SISD: the conventional Von Neumann architecture,
SIMD: the array processors,
MISD: the pipelined processors,
MIMD: the shared memory multiprocessors.

In array processors several independent identical arithmetic units (ALU) or processing elements (PE) operate in parallel under supervision of one Control Unit (CU) (Fig. 3).

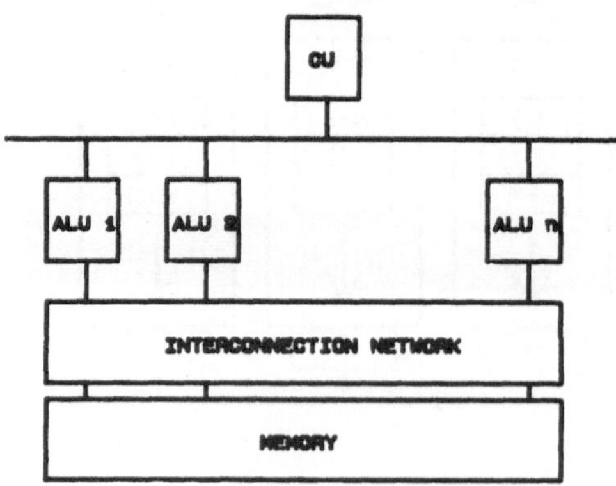

Fig. 3. SIMD Array Processor Architecture.

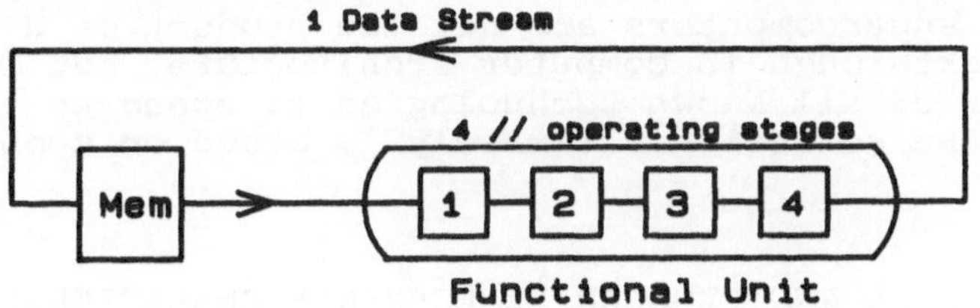

Fig. 4. MISD Pipelined Architecture.

Conversely, in pipeline computers, a single datum is operated upon by the different stages of one Functional Unit (FU). Parallelism is achieved by the simultaneous execution of the different stages on a stream of sequential data (Fig. 4).

In the two preceding techniques, the execution time of an elementary operation of a Processing Element or a Pipeline Stage is one time unit. This facilitates the synchronisation between processing elements, since all PE's operate in lockstep under control of the same clock.

Multiprocessors on the other hand allow different processors to execute parts of a program asynchronously. While this allows a greater flexibility, such an architecture also requires a fast synchronisation system for the asynchronously operating PE's. This also greatly enlarges the complexity and puts an additional overhead on a fast parallel system (Fig. 5). This is probably the main reason why today's supercomputer do not use multiprocessing techniques in general.

From the previous description it is evident that there is no single answer to the need for faster processors. There are a number of techniques, and the computer architects have realized machines with a great peak performance only from a blend of the five techniques:

- fast processor and memory technology
- wide high-speed databus
- multiple pipelining
- lockstep operation
- parallel processing

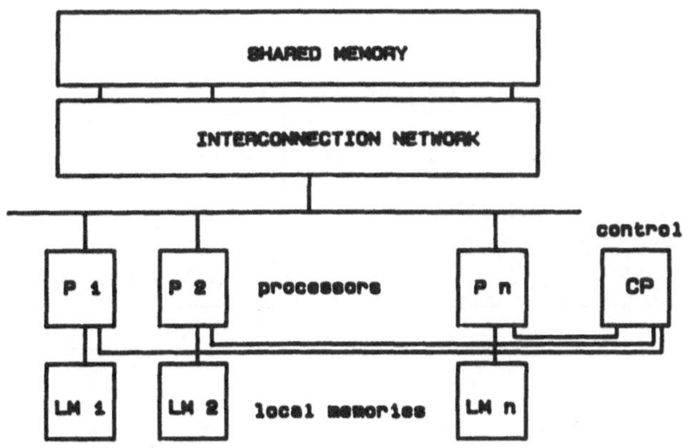

Fig. 5. MIMD Multiprocessor Architecture.

Clearly, supercomputers are not the product of a nicely developed breakthrough in computer architecture, but incorporate a balanced mix of all known technologies to speed-up the processing power, whose selection ultimately is based on economical trade-offs.

III. ARCHITECTURAL AND PROGRAM PERFORMANCE CHARACTERIZATION

The combination of the above techniques makes the performance evaluation more difficult. While the brute force methods like fast processor and memory technology deliver a uniform speed-up for all processor-bound calculations, the parallel techniques such as pipelining and array processing are strongly dependent on the program for their effect on the performance. We will therefore discuss briefly the known performance measures for these architectures [3].

3.1. Pipelining

All available supercomputers use pipelining techniques for vector as well as for scalar operations. Pipelining applies to a complex processor function Y = F(X), when the following conditions are met:

- the execution time of the function F is a multiple of the time τ to fetch a single argument;
- the function can be decomposed into a number of sequential operations, organized as hardware stages and implementing the function F as F = f1(f2(... fl(X)...)), such that each stage is able to operate independently on the results of the previous stage;
- the execution time of each stage equals τ.

Under these conditions the function F can be organized as a pipe, consisting of l stages, each stage operating as in an assembly line (Fig. 6), on consecutive data.

The success of pipelining - the prime contributor to a peak processing rate - depends on

- the matching of pipeline and datastream speeds;
- the data-alignment.

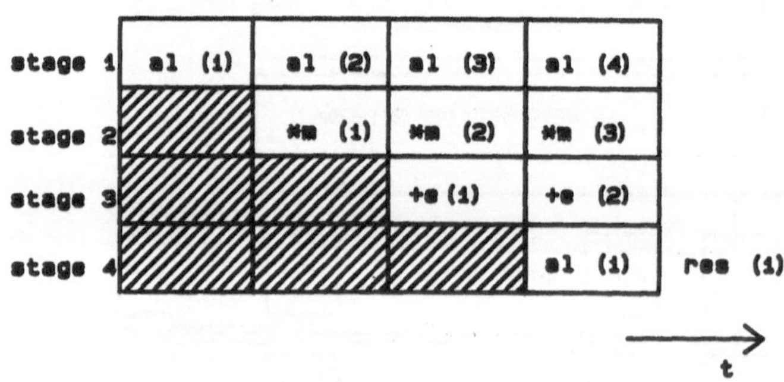

Fig. 6. Pipeline schedule for multiplication. Al = alignment of mantissas, *m = multiplication of mantissas, +e = addition of exponents.

Table I. Pipeline parameters for scalar multiplication in three machines.

	CRAY-1	CYBER-205	FPS AP-120B
stage-time	12.5 ns	20 ns	167 ns
# stages	7	5	3
min. fetch	12.5 ns	20 ns	167 ns
max. fetch	50 ns	380 ns	333 ns

a) Pipeline matching. A chain is as strong as its weakest link, and so is the pipelined processing element. The links of a pipe are the stages and the datapath to and from the memory. As noted earlier, memories are organized in interleaved banks, yielding a minimum data access time τ. In order to match this datarate, the pipe is designed as a number of stages, each of which has a processing time $\tau' < \tau$. This pipe matching typically occurs in the CRAY and Cyber computers where pipelining is applied to different "Functional Units"; a sample of the pipe parameters for the scalar multiplication on these machines is shown in Table I. From this table it appears that the weakest link is the datapath. The minimal data access times only apply when the data are fetched from different memory banks (minimum 4 banks for the CRAY-1, 2 for the FPS), or when the special datapipe 'load/store' is empty in the case of the Cyber 205.

b) Data alignment. The critical effect of the data-access on performance is enhanced by the difficulty to organize the data in a sequence which ensures minimal access times. Most memories used in pipelined machines are organized into an n-way interleaved memory, in which each bank has an access time of $c = \tau.n$. When data are addressed in consecutive banks, this leads to an interleaved access time τ. This occurs for a sequential vector access when the vector elements are stored in consecutive banks. However, when the consecutive vector elements are separated by an interval p, the real access time of each element becomes:

$$a = \tau.\gcd(p,n) \qquad (1)$$

where gcd(p,n) is the greatest common divisor of p and n. This is illustrated in Fig. 7 for an 8-bank memory.

Although the average access time for all possible intervals is close to the clock periods for up to 8 memory banks, real combinations of p and n produce typical access patterns with much larger access times. This is because the power of 2 naturally arises in both the hardware and software. At the hardware side, the number of banks usually is a power 2, because this simplifies the address-logic. This creates memory conflicts, e.g. when a column-wise stored matrix is accessed by row, and the dimensions of the matrix are a multiple of the number of banks. Since many frequently used algorithms are also based on powers of 2, such as the FFT and 'recursive doubling'

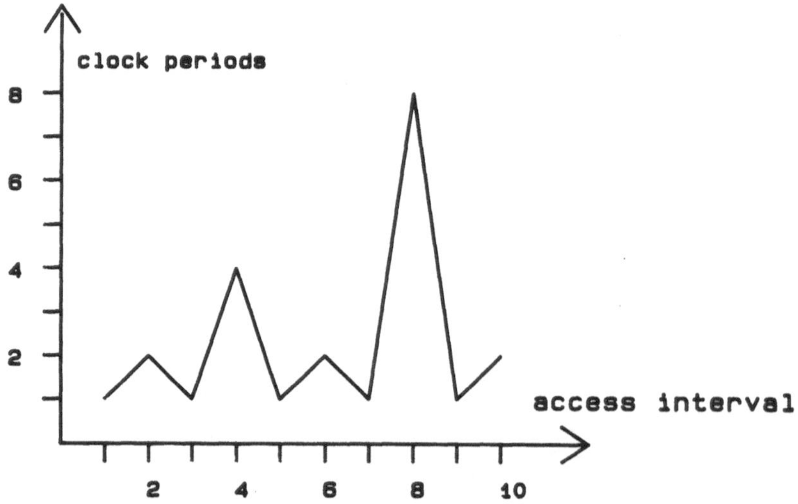

Fig. 7. Access time of p-spaced vector access in an 8-bank memory.

techniques (divide and conquer), this type of conflicts is not an exception.

In order to reduce this type of memory conflict, there are two alternatives:

a) use a prime number of memories. Selecting a prime number of banks will eliminate access delays for all equally spaced access patterns; therefore a prime number of memory banks gives a much better overall access time [4].
b) use prime array dimensions. This is a software solution with the same result, but requires careful programming and sometimes redundant memory reservation for fast algorithms based on powers of two.

It is interesting to note the similarity between a pipelined functional unit and an interleaved memory: the memory can be considered as a functional unit, producing results (data) at each clock period, delivered by stages which are memory banks.

c) Pipes and performance. Since a functional unit organized as a pipe processes a data stream at one result per clock period τ, it is worthwhile to build a very powerful unit. However, the maximal performance is attained only when the pipe is full, i.e. there has to be a continuous datastream, producing one set of arguments for each clock period. An empty pipe with l stages calculates n results in a time

$$t(n) = (l + n - 1)\tau.$$

Using the half-performance length $n_{1/2}$, defined by Hockney and Jesshope [3] as the vector length for which the execution time equals 2τ per vector element, one finds

$$t(n_{1/2}) = 2.n_{1/2}.\tau = (l + n_{1/2} - 1)\tau$$

or

$$n_{1/2} = l - 1 \quad (1)$$

In other words, a 50% performance necessitates that the pipe is fed with batches of at least l - 1 operations, e.g. 2 - 6 for the multipliers of Table I.

d) Linked Pipes. Most supercomputers have 2 (FPS) to 12 (CRAY-1) pipelined functional units (FU's). For some calculations the performance can be doubled or tripled, if several PU's are able to operate together. While it is impossible for the architectures of Table I to produce a double or triple data rate from the memory, the operation is feasible for two consecutive different floating point operations on a single data stream, e.g. in recurrence relations or the Borner polynomial evaluation scheme, with calculations of the type:

$$d_i = a_i + b_i * c_i.$$

This requires that the two FU's (* and + in this case), can be linked together forming one long pipe (Fig. 8). The corresponding operations are called 'triadic instructions'. Provisions in both the CRAY-1 and Cyber 205 are made for that purpose, respectively named "chaining" and "shortstopping". It should be noted that this linking increases the half-performance length due to the length of the concatenated pipes, l_1 and l_2, and possibly a couple of intermediate stages l_i, leading to

$$n_{1/2} = l_1 + l_2 + l_i - 1. \quad (3)$$

3.2. Array Processing

Whereas pipelining refers to a single multistage functional unit, array processing refers to an array of identical processors, all performing identical operations at the same time and under centralized control.

An array processor is an SIMD machine, consisting of arithmetic and logical units (ALU's), which execute the same instruction under supervision of one control unit (CU). Data access is through one or several memories, using multiple data paths. Because generally the ALU's are not as fast as the stages of a pipeline, there is no demand for fast sequential data access. In contrast, many memory locations should be

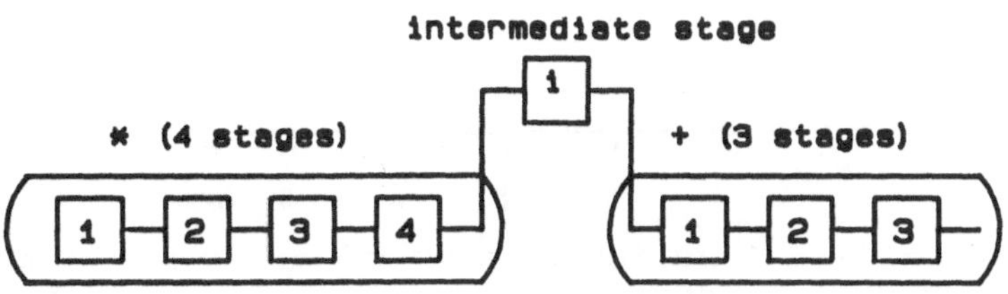

Fig. 8. Linking Pipes.

available at the same time for storing and fetching the arguments of each ALU. Therefore the memory is organized into many concurrently accessible banks, which are connected to any of the processors by a suitable interconnection network. By its conceptually different architecture, array processors have other bottlenecks than pipelines in achieving a high performance, these are:

- the need to keep all ALU's busy;
- the interconnection and routing delays.

a) array processing and array programs. Array processors are conceived for array computations, preferably on vectors with lengths equal to a multiple of the number of ALU's. The processing rate follows a sawtooth function in the vector length, as is shown in Fig. 9.

With p processing elements, the relative speed-up S(n) for a vector operation on n elements is:

$$S(n) = (n/p) \,/\, \lceil n/p \rceil$$

with $\lceil x \rceil$ the largest integer smaller than x+1. This gives

$$n_{1/2} = p$$

for a minimum 50% performance in all cases.

Of course, not all program statements are vector operations and this can seriously handicap a sustained parallel operation. Consider a program run on an array processor with an arbitrary number of processing elements (PE's). At any moment during the computation there are w(t) parallel operations. Clearly, the average number of busy processors equals the average

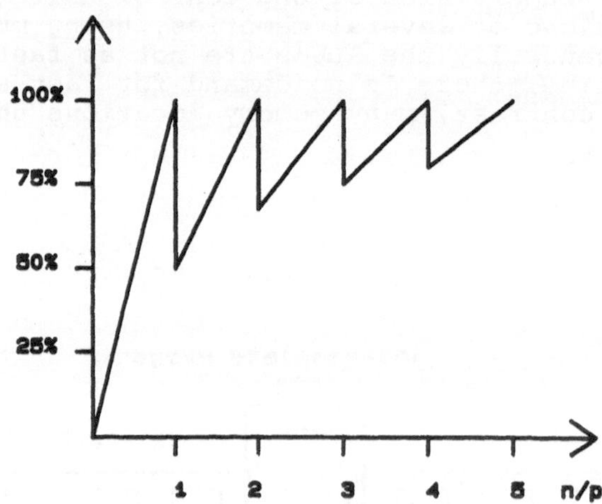

Fig. 9. Normalized fraction of maximal array performance with vector length n and number of processing elements p.

parallelism W of the program:

$$W = 1/L \int_0^L w(t)\, dt$$

The total work done by all processors equals its execution time E on a uniprocessor and one has

$$E = \int_0^L w(t)\, dt$$

With t_p representing the execution time using p processors, the maximal speed-up is $S = t_1/t_{inf}$. Here $t_1 = E$ and $t_{inf} = L$ and one has

$$S_{max} = E/L = W, \tag{4}$$

so the maximal speed-up equals the average parallelism in the program.

b) Distribution of parallelism. Now different families of programs can be defined, depending on the probability distribution of w(t). Therefore assume a p processor operation and define q_i as the fraction of the work that is done at degree of parallelism i. Then one has:

$$t_p = (q_1 + q_2/2 + \ldots + q_p/p)\, E \tag{5}$$

and

$$S_p = t_1/t_p$$

$$= 1/(q_1 + q_2/2 + \ldots + q_p/p). \tag{6}$$

Now different hypotheses on the values of q_i can be worked out [5].

- Equal work hypothesis: Assume an equal amount of work is done at all degrees of parallelism, i.e. $q_1 = q_2 = \ldots = q_p = 1/p$ (Fig. 10).

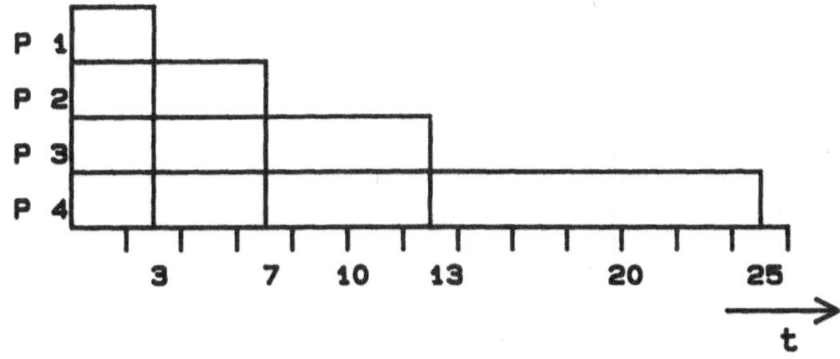

Fig. 10. Equal work schedule. Total work is 48 time units.

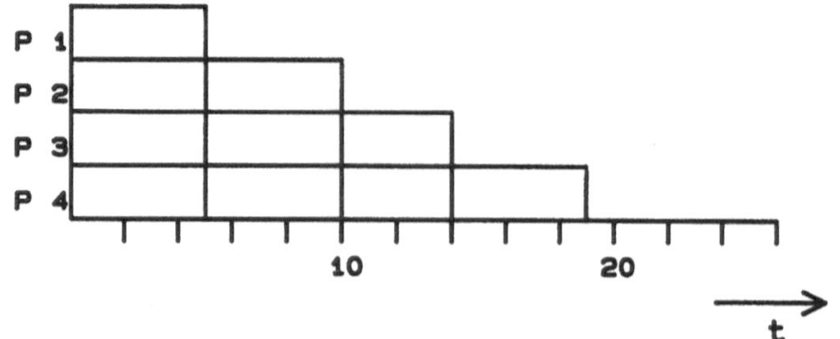

Fig. 11. Equal time schedule with the same workload as in Fig. 10.

Then substitution of q_i into (2) gives

$$Sp = p(1 + 1/2 + \ldots + 1/p)$$

Now one has, with Hp the p-th harmonic number

$$Hp = 1 + 1/2 + \ldots + 1/p > \ln(p)$$

which leads to

$$Sp = p/Hp < p/\ln(p) \qquad (7)$$

- Equal time hypothesis: Here one assumes that the execution time at each degree of parallelism is equal, i.e. $q_1/1 = q_2/2 = \ldots = q_p/p = a$, or else $q_i = i.a$ (Fig. 11). Since $q_1 + q_2 + \ldots + q_p = 1$, one finds $a = 2/(p(p+1))$ and eq. (6) yields

$$Sp = 1/(p.a) = (p+1)/2 \qquad (8)$$

- Inverse work hypothesis: The fraction of work to be done at each degree of parallelism is assumed to be inversely proportional to the degree of parallelism, i.e. $q_i = K/i$ (Fig. 12).
With

$$S\ qi = K\ S(1/i) = K.Hp,$$

one has

$$K = 1/Hp < 1/\ln(p)\ .$$

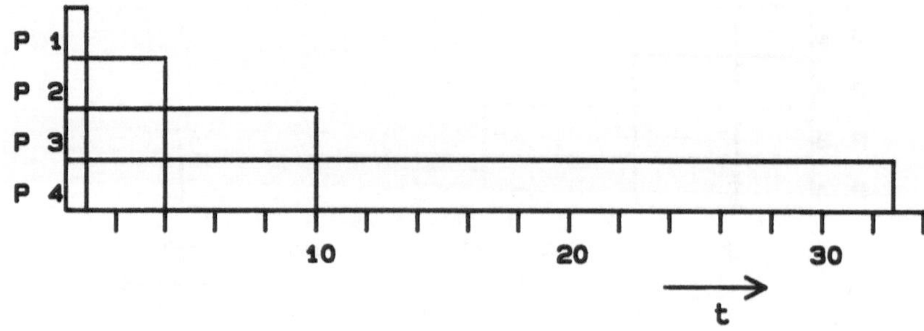

Fig. 12. Inverse work schedule with workload 48 time units.

Furthermore, eq. (6) becomes

$$Sp = Hp/(1 + 1/2^2 + \ldots + 1/p^2) = Hp/\sum_{i=1}^{p} (1/i^2)$$

and with

$$\sum_{i=1}^{p} (1/i^2) < \sum_{i=1}^{\infty} (1/i^2) = \pi^2/6$$

one finds

$$Sp > 6.\ln(p)/\pi^2 \tag{9}$$

c) Empirical considerations. Lee's empirical observations on a large number of programs, using the Parafrase Analyser of the University of Illinois, revealed real speed-ups to lie between the upper and lower bounds of the three previous hypotheses. The majority of computations have a speed-up

$$\max(1, k1.\ln(p)) < Sp < \min(k2.p/\ln(p), p)$$

The speed-up of the programs is correlated with the Lee's approximation formulas, as appears from Table II.
It is important to note that the utilisation of the processors and therefore its economical efficiency drops as the number of processors increases. Consequently, general use of a massive number of processors is not likely to break through, unless VLSI techniques allow to reduce significantly the cost per processor.

d) Worst case speed-up bound. One of the difficulties in using Lee's hypothesis is that the characterization of the q-distribution requires a thorough analysis of the program structure. We present an alternative estimate based on a worst case analysis, taking into account the number of processors and the "longest path" in the program [6]. The longest sequence of calculations which must be carried out sequentially due to precedence constraints, is called the longest path and by its nature must be calculated on one processor, thereby setting a minimal execution time on the program execution. In the best case, all other calculations can be done in parallel with the longest path, yielding a speed-up Sp = E/L where E is the total amount of work in the program, or its execution time on one processor. In the worst case, no calculation can be carried out in parallel with the longest path and the remaining work E - L has to start after finishing the longest path tasks. In order to be consistent with the definition of longest path, this work has to be perfectly partitionable on an infinite number of

Table II. Empirical speed-up observations for a large number of programs.

Range of processors	Order of speed-up
1-10	Sp = 0(p)
11-1000	Sp = 0(p/ln(p))
1001-10000	0(ln(p)) < Sp < 0(p/ln(p))
> 10000	Sp = 0(ln(p))

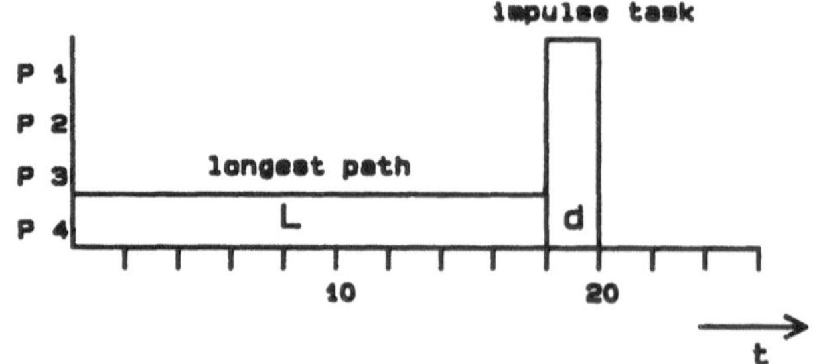

Fig. 13. Worst case impulse task schedule.

processors p, giving a zero additional execution time for an infinite number of processors. Therefore the work E - L is called an "impulse task" (Fig. 13). On a finite number of processors, the impulse task causes an extra execution time of e = (E-L)/p, leading to a worst case execution time

$$t_p = L + e = (E + L(p-1))/p$$

and a worst case speed-up

$$Sp = t_1/t_p = p/[1 + (p-1)\ L/E].$$

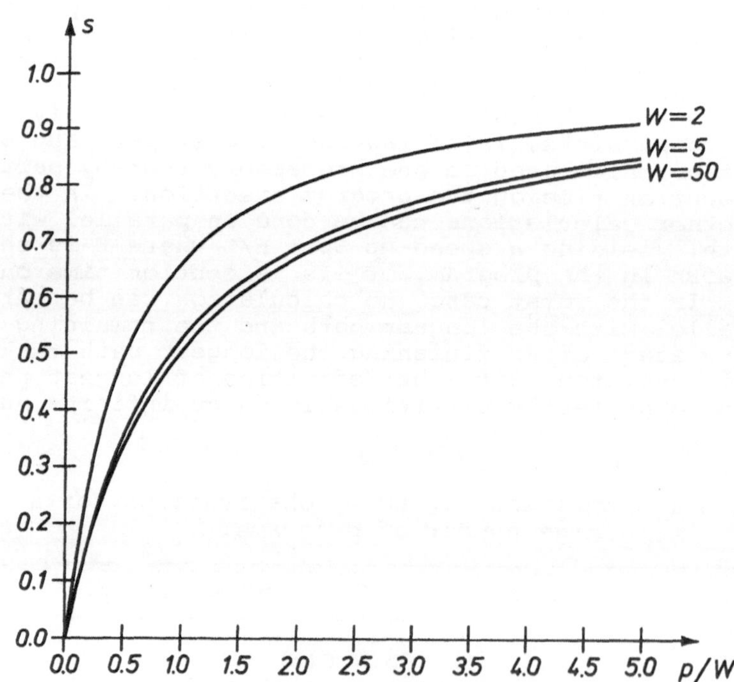

Fig. 14. Maximum relative speed-ups, for worst case program. p = number of processors, W = average program parallelism.

Recalling that W = E/L is the average parallelism in the program one finds

$$Sp = p/[1 + (p-1)/W] \tag{10}$$

and with Sp,max = E/L = W the relative speed-up s is defined:

$$s = Sp/Sp,max = p/W/[1 + (p-1)/W] \tag{11}$$

which is depicted in Fig. 14.
The half performance number of processors $p_{1/2}$, needed for a 50% of the maximal speed-up W follows from (11) with Sp/Sp,max = 1/2, giving

$$p_{1/2} = W - 1 \tag{12}$$

In Lee's terms, this worst case analysis corresponds with a q-distribution

$$q_1 = L/E = 1/W$$

$$q_2 = q_3 = \ldots = q_{p-1} = 0$$

$$q_p = (E-L)/E = 1 - 1/W$$

Finally, the equal work hypothesis is intuitively appealing for parallel algorithms using a 'divide and conquer' approach, such as the recursive doubling technique, where the number of active processors in each time step is divided by 2. E.g. the execution time for calculating the sum of 2^{n+1} elements using $p = 2^n$ processors equals p, giving a speed-up

$$S = 2^n/n = p/\log_2(p)$$

which corresponds to Lee's equal work hypothesis.

e) Data routing. Array processors require a new set of p arguments for every calculation step. For efficient operation, these arguments have to be fetched simultaneously and this requires a large number of m parallel accessible memory banks. Processors and memories are connected at best by a fast pxm cross-bar switch, which allows common access to all memories.

With a memory bandwidth of w bits, a cross-bar contains $p \times m \times w$ switches, which becomes economically unfeasible for moderate p and m. Moreover, memory conflicts still occur when multiple arguments are stored in the same memory. On the other hand, only a fraction of the N^M possible cross-bar switch settings is actually used. More economical switches include multistage networks, but they require an additional routing delay for each stage.

Routing delays of cross-bar and multistage networks can be minimized by overlapping the data stream with the execution stream. This kind of pipelining has led to a rather complex alignment network in the Burroughs scientific processor, which contains 16 ALU's and 17 memory banks. Essentially in this computer the execution stream is matched with the data stream by a number of microcoded templates. The hardware router will predict the free slots in the data fetch pipeline and select the best possible microcoded execution sequence from a set of precoded alternatives. Each alternative provides another sequence of basically the same instruction stream. The router will combine different templates in order to minimize the number of free slots in the interconnection network.

3.3. Pipelined versus Array Processing Architecture

The principal items of comparison are shown in Table III. From this table it appears that theoretically array processors with a large number of PE's should be able to outperform the pipelined architectures. However, routing delays and absence of parallelism in the average program causes an array processor to be a very specialized machine.

A pipelined supercomputer on the other hand usually consists of a number of high speed functional units. This is necessary in order to provide a well balanced calculation speed for scalar and vector operations on integer, floating point and character data. As a result, architects of popular supercomputers tend to apply primarily pipelining techniques and array processing often implemented as an add-on enhancement by replicating the existing pipes a number of times, such as an up to four pipes Cyber 205.

A principal bottleneck of array processors is the need for identical parallel operations. More flexibility is gained when each processing element can process a different instruction stream, i.e. is operating as an MIMD multiprocessor. While more parallelism can be extracted from the program, new problems such as synchronization and memory conflicts arise.

IV. DYNAMIC PERFORMANCE MODELS

Generally supercomputers have a wide performance range, when performance is measured by the Megaflop rate. Top performance is only obtained through fine tuning of the program to the

Table III. Elements of comparison between pipeline and array processors.

	pipeline	array
Independent Unit (IU)	stage	processing element
general purpose IU?	NO	YES
number of IU's	2-7	16-4096
half perf. length	#IU's-1	#IU's
speed of one IU	20-50 ns	320 ns - 250 μs
fast scalar ops.?	YES	NO
speed-up by linking?	YES	YES
memory access	interleaved	network
memory delay	start up delay	router delay
programming effort	moderate	high
program dependent	moderate	high

Table IV. Correspondence between queueing and computer terms.

Queueing term	Computer term
Service station	Processors, Routers, Memory banks
Server	Processor, Router, Memory
Customer	Operation, Task
Class of customers	Scalar fetch, Scalar operation, Vector fetch, Vector operation

hardware architecture. We already invoked the impact of the available parallelism in the program on the maximal speed-up. In this section, emphasis is put on the interaction between the different resources of the architecture, such as the scalar and vector pipelines, the memory banks and the routing network. For that purpose, simple queueing models are presented, in which the resources are presented as "servers", delivering a "service" to "customers". Each "service station" serves various "classes of customers", characterized by their different service time and routing probability . Service times are taken from an exponential distribution with a class-dependent average. Resources usually deal with one "request" at a time, on a "first come, first serve" (FCPS) basis. However, a resource can have two or more servers, accepting requests up to the number of servers available. Not serviceable requests wait in a "queue" associated with the resource.

Customers go from one resource i to another resource j with predefined probabilities p_{ij}, which are stored in a transition matrix T. The set of all classes and corresponding resources which can be visited by a customer is called a chain. All chains together form a queueing network. Here we deal with closed queueing networks, i.e. the customers stay in the network and the number of customers remains constant. Furthermore there is only one chain, i.e. each customer can visit any service station. Table IV represents the queueing network terminology and the corresponding computer terms.

These models are solved by a queueing networks package 'QNET' developed at the LEM-laboratory. The system accepts a high level description of the queueing network and produces numerical and graphical output in an interactive session at the terminal. Both parameters and model structure are easily changed and the model analyst can experiment interactively with different possible alternatives. The solution methods implemented are the Local Balance and the Mean Value Analysis algorithms [7,8].

4.1. Pipelined Services

Pipelining requires special attention in queueing networks, because the service time varies with the number of customers. A pipeline is characterized by a start up time and a steady state time, respectively $l.\tau$ and τ where l is the number of pipeline stages and τ is the stage time. A given

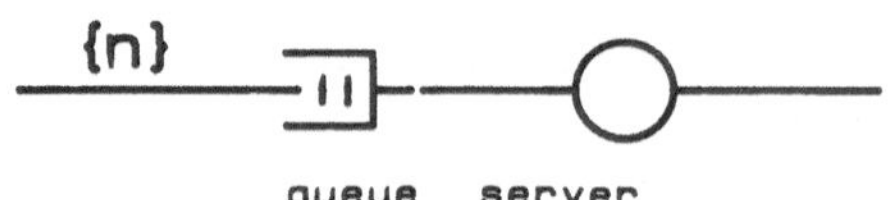

Fig. 15. Pipeline queue.

number of n customers arriving in the pipeline will receive service in a time (n+l-1)τ. Therefore we model a pipelined server as a sequential server, where a group of n requests is considered as one customer, with a varying service time (Fig. 15).

4.2. Chaining Pipes

In the CRAY-1 and Cyber 205, two pipelined units with lengths l_1 and l_2 can be chained together to form 1 pipeline with length $l_1 + l_2$. Chaining therefore can be modelled as a single server with variable service time $(n+l_1+l_2-1).\tau$ for n requests.

4.3. Processor - Memory Pipes

Processor and memory queues both operating in a pipelined fashion cannot be joined into one queue, because they operate independently. Nevertheless, processor and memory can momentarily operate as one pipeline, typically on a non pre-emptive vector operation which is programmed as a burst of pipelined requests. When the combined processor-memory pipe is full, customers leave the processor and memory pipe at the rate of the slowest server. In order to account for the fixed start up time of l clock periods, the processor service time is set to lτ. After the start up period, processor and memory operation overlaps, and the pipelined request for N operations is programmed as follows (assume $\tau_m > \tau_p$):

- memory service time $\tau'_m = N.\tau_m$ (13)
- processor service time $\tau'_p = l.\tau$

V. QUEUEING MODELS OF SUPERCOMPUTERS

5.1. CRAY-1

A queueing model for the CRAY-1 is given in Fig. 16.

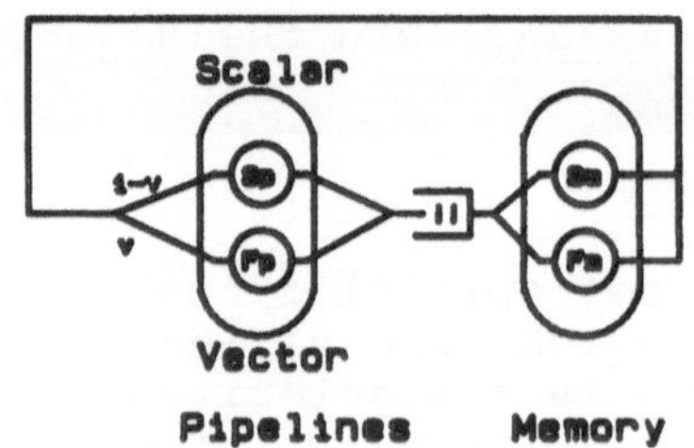

Fig. 16. CRAY-1 queueing model.

It consists of a scalar and a vector unit, operating in parallel. Both units have access to the same multibank memory which has an interleaved cycle time of 12.5 ns for vector data and 25 ns for scalar data. The Megaflop rate is calculated as a function of the ratio of vector calculations, v, and the average vector length N.

In the CRAY-1 the floating point processors are shared between scalar and vector units. Therefore the queueing network consists only of two queues: a processor unit and the memory. Both queues have a <u>scalar class</u> (Sp, Sm) and a <u>vector class</u> (Fp, Fm) of customers. The average service times are taken from the multiplication functional unit:

Spt = 87.5 ns: start up time for (1-v)N scalar operations
Smt = 25 ns: memory cycle time for 1 result
Fpt = 87.5 ns: start up time for vN vector operations
Fmt = 12.5 ns: memory cycle time for 1 result

The vector memory cycle time is smaller, because vectors have a fast intermediate store of 64 elements, implemented as a "register file". The processor times Spt and Fpt are calculated according to eq. (13). Spt equals the start up time for (1-v)N multiplications, i.e. the time required to calculate the first result; Fpt represents the start up time for a burst of v.N multiplications. Smt and Fmt are the memory service times for a single scalar or vector result. Since one customer represents a burst of vN or (1-v)N operations, the actual memory service time is vN.Fmt for vector or (1-v)N.Smt for scalar customers.

The processing power is measured as the number of customers served per second for both classes, Sp and Fp. Since one Fp customer represents v.N floating point operations, the processing power in Megaflops is expressed as

$$P_{CRAY-1} = (RSp.(1-v) + RFp.v.Nops).N$$

where RSp and RFp are the service rate of the scalar and vector customers. N = the vector length and Nops = the number of calculations on the same arguments. For diadic operations Nops = 1, for triadic operations Nops = 2. P_{CRAY-1} is plotted against v for various values of N and Nops in Fig. 17.

From these figures it appears that for triadic operations a vector-performance (v = 1) of more than 40 Mflops is achieved for vectors of less than 5 elements. The pipelined vector unit is most beneficial for triadic operations, involving two <u>different</u> floating point units, in which two pipes (addition and multiplication) can be chained together to form one pipe, thereby doubling the vector performance. For short vectors, a minimum in the performance is observed. This is caused by the double start up times for the serial and the vector calculations, when $v \neq 0$, which are relatively important for small vectors.

5.2. CDC Cyber 205

In contrast with the previous architecture, this computer has an independent pipelined scalar unit, performing addition and multiplication a with start up time of 100 ns and steady state rate of one result in every 20 ns. The separate vector unit contains one to four floating-point pipes. In each pipe the addition and multiplication units are able to operate at a 50 Mflop rate.

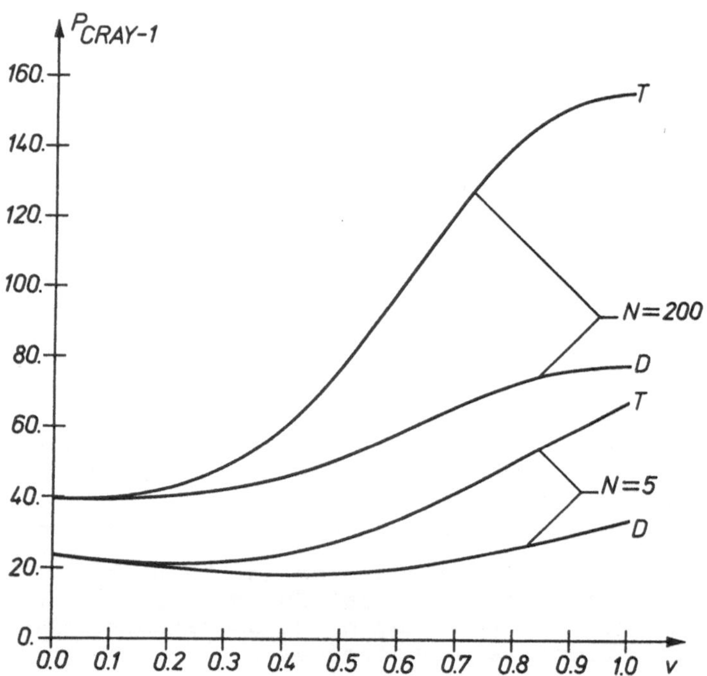

Fig. 17. CRAY-1 scalar versus vector processing power. v = fraction of vector computations, P = performance in Megaflops, N = number of vector elements. D = diadic operations $c_i = a_i * b_i$; T = triadic operations $d_i = a_i + b_i * c_i$. All operations are in 64-bit arithmetic.

The scalar unit accesses the main memory through a memory interface unit, which can be considered as a fast cache memory, delivering data at 20 ns per argument. The interface unit communicates with the multibank main memory through a 512 bit databus at a rate of one 512 bit word per 20 ns. These extra long words are stored into intermediate read or write buffers.

The vector unit is communicating directly with the main memory in 2 input streams and one output streams, each delivering 128 bits data in 10 ns.

A queueing model of the Cyber 205 therefore consists of 4 service stations: the scalar unit, the vector unit, the memory interface unit and the main memory unit (Fig. 18).

Again, the performance of a mixed scalar/vector operation is considered, where v is the ratio of the vector over the total number of operations and N equals the average vector length. Start up times for scalar and vector units are 100 ns and 1.040 μs respectively. The steady state data rate of main

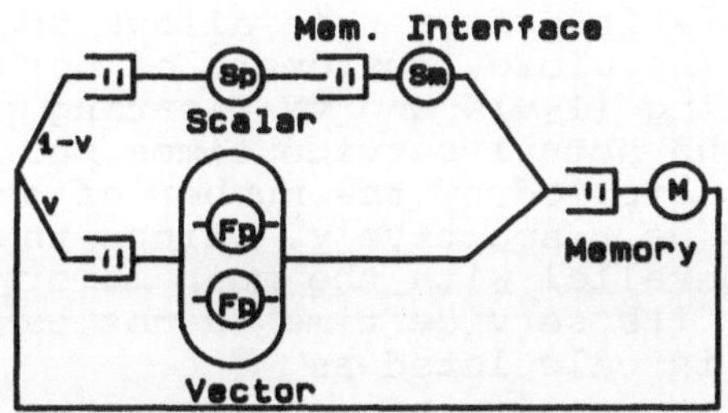

Fig. 18. Cyber 205 queueing model.

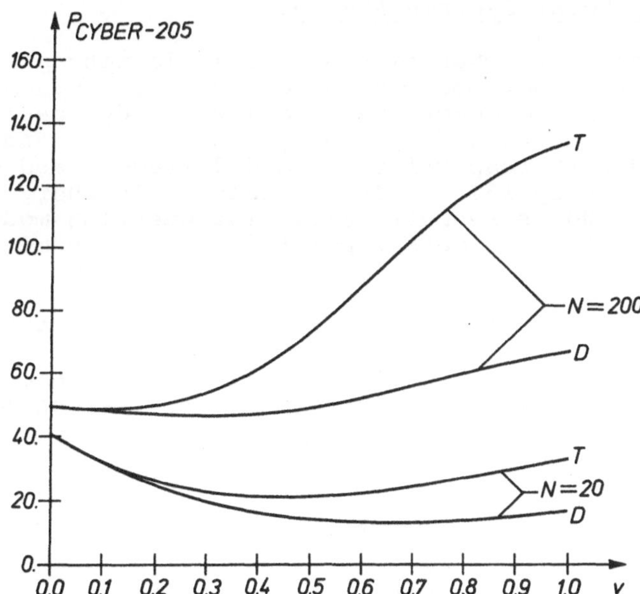

Fig. 19. Processing rate of Cyber 205. v = fraction of vector calculations, P = performance in Megaflops, N = vector length, p = number of pipes, D = diadic, T = triadic operation. All calculations are in 64-bit arithmetic.

memory and memory interface unit both equal 20 ns per operation. The model assumes 'npipe' floating point pipes, sharing the vector calculations. The service times are:

Spt = 100 ns: start up time for (1-v)N scalar pipelined operations
Smt = 20 ns: memory interface time per pipelined operation
SMt = 20 ns: main memory scalar service time per operation
Fpt =1040 ns: start up time for vN vector operations
FMt = 10 ns: main memory vector service time per operation

Note that the memory interface unit allows the scalar operations to be pipelined. Therefore, customers of both scalar and vector type are representing (1-v)N and vN floating point operations respectively and the actual service times per customer, SMt and FMt, have to be multiplied by the number of operations, giving 20(1-v)N and 20 vN ns respectively. Since the memory interface unit operates in parallel with the main memory and has the same service time, its service time is not taken into account. The Megaflop rate is calculated as

$$P_{\text{Cyber 205}} = (\text{RSp.}(1-v) + \text{RFp.v.npipes.Nops})N$$

and is shown in Fig. 19.

In contrast to the CRAY-1, the maximal performance for small vectors decreases when using the vector unit. This is due to the long start up time of the vector pipe, which requires a large vector length to compensate for. Consequently, the Cyber 205 becomes most efficient only for very large vectors. The asymptotic performance of a 1-pipe system on triadic operations is 200 Mflops.

5.3. Floating Point Systems AP-120B

This number cruncher consists of a floating point adder and multiplier, programmed by large microcode sequences. Interleaved access in two memory banks allows a best case cycle time of 167 ns, using a fast memory. Both adder and multiplier are pipelined with respectively 2 and 3 stages, and can deliver one result in every 167 ns after a relatively short start up of 333 ns and 500 ns respectively. The queueing model therefore consists of the floating point unit and the memory (Fig. 20).

The effective performance depends on the vector length N and on the number r of memory references per operation. R is 3 for a memory to memory operation, making the memory the bottleneck which limits the performance to 2MFlops (500 ns/

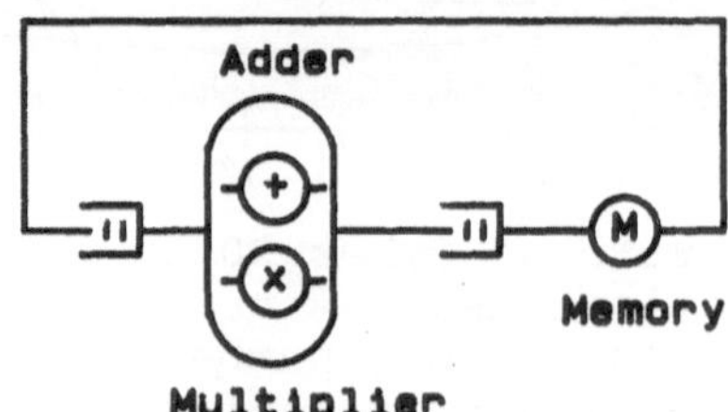

Fig. 20. FPS AP-120B queueing model.

result) for this type of operations. However, for some algorithms such as matrix multiplication, careful programming allows a value of r = .66/n, with n the matrix dimension. In this case, the intermediate results are fed back into the pipeline and do not require a long memory transfer. This allows a maximum processing rate of 6 MFlops, for triadic operations, yielding one addition and one multiplication per clock period.

The parameters of the FPS model are

Spmult = 500 ns: the start up time of the multiplier
Spaddt = 333 ns: the start up time of the adder
Smt = 167 ns: the interleaved memory access time

With these parameters the performance ranges between 1 and 6 MFlops (Fig. 21).

From this figure, the major influence of the memory access time is apparent. As a consequence, optimal use of the Floating Point Systems AP-120B requires a careful assembler programming in order to streamline the data flow to and from the memory. The highest performance is obtained for structured

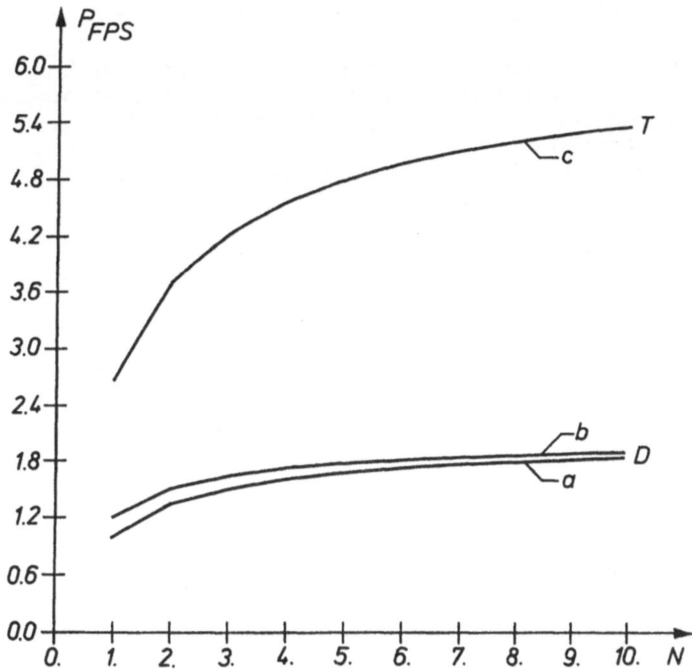

Fig. 21. FPS AP-120B performance. N = the number of pipelined vector operations; a) memory to memory vector multiplication; b) memory to memory addition; c) triadic operations of the type $S = S + a_i * b_i$.

calculations such as the Fast Fourier transform or matrix multiplication.

5.4. Burroughs Scientific Processor (BSP)

Although not commercially available, an analysis of this computer is worthwhile, because of its array processor architecture. There are 16 ALU's operating under control of one master unit, which communicates via a store and fetch routing network directly with a 17 bank data memory. Consequently we have here a real parallel processor, of which the queueing model is given in Fig. 22.

All ALU routing and memory operations are pipelined with a relatively slow clock period of 160 ns. The effective processing rate depends on the slowest stage in the pipe. The execution time of every stage is:

ALU : 320 ns for multiplication or addition
Routers: 160 ns per argument
Memory : 160 ns per argument-fetch or -store

Since Fetch and Store Router operate in parallel, only the service time of the Fetch Router is taken into account. Also memory and ALU operations are overlapped, consequently only the longest of the two service times is required, which corresponds to a memory bound or a processor bound calculation. Since an arithmetic operation requires two clock periods, the cross-over point is reached for 1 ALU operation per 2 memory cycles. This average is realized for triadic operations of the type $d_i = a_i + b_i * c_i$, where 2 operations are done on 4 arguments (3 fetch, 1 store), i.e. 0.5 operations/argument.

The effective performance finally depends on the ALU-occupancy, i.e. the obtained figures for a 16-ALU operation have to be corrected by the reduction factor of fig. 9. In Fig. 23 the performance range is given for a full vector operation (16 ALU's), for calculations in bursts of 2 to 10 consecutive vector elements.

The performance in Megaflops is

$$P_{BSP} = RALU * Nops * 16$$

with RALU the number of customers per second visiting one ALU

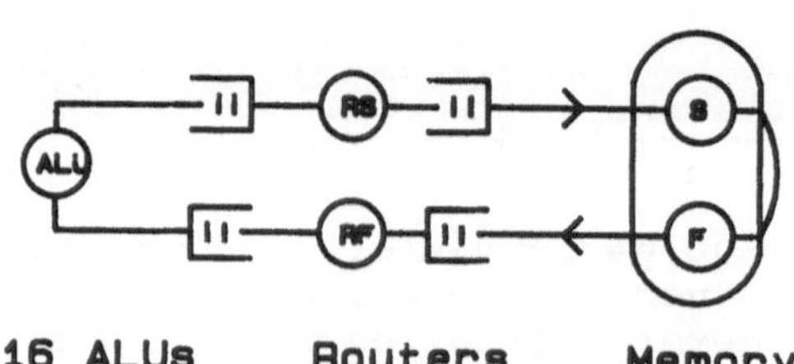

Fig. 22. BSP queueing model.

and Nops the number of operations done by the ALU on one set of arguments.

Clearly, the performance increases by about 50% for triadic operations, in contrast to the memory-bound diadic operations. Although the half-performance length for each ALU is less than 2, the effective half-performance length is 16 times this value, i.e. between 20 and 30.

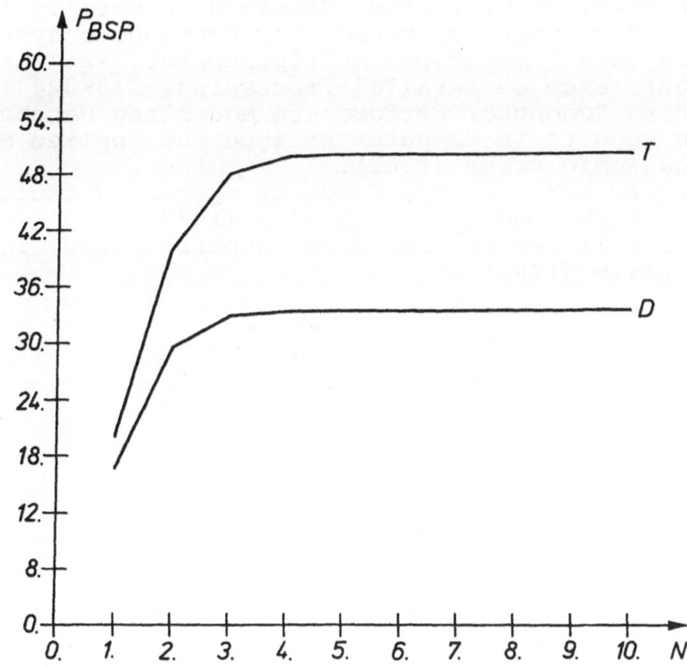

Fig. 23. Maximal performance of the BSP for full 16 ALU pipelined vector calculations on 2 to 10 elements per ALU. D = diadic operations; T = triadic operations.

VI. CONCLUSION

Supercomputers are very fast machines which, according to today's standards, have an enormous computing power. This performance is not free; in order to create such a machine the combined effort of high speed technology and new architectural design is needed. In order to extract its virtual power, a careful study of the machine and the program is required. In this paper the different factors which determine the ultimate performance were discussed and combined in global performance evaluation models.

REFERENCES

1. M.C. Vanwormhoudt, Efficient data handling using multiprocessor structures, in: Flanders Technology International Seminar: "Parallel Information Processing and Artificial Intelligence in Simulation", 21-39, 28 Feb. 1985.
2. M.J. Flynn, Some computer organizations and their effectiveness, IEEE Transactions on Computers, 21:948-960 (1972).
3. R.W. Hockney, and C.R. Jesshope, "Parallel Computers", 416 pages, Adam Hilger Ltd. (1981).
4. D.H. Lawrie, The prime memory system for array access, IEEE Transactions on Computers, 31:435-442 (1982).
5. R.B. Lee, "Performance Characterization of Parallel Computer Organizations", Ph.D. Thesis, Stanford University (1980).
6. E.H. D'Hollander, Speed-up bounds for continuous system simulation on a homogeneous multiprocessor, in: International Conference on Parallel Processing, 318-324 (1981).
7. S.S. Lavenberg,"Computer Performance Modelling Handbook", Notes and Reports in Computer Science and Applied Mathematics, Academic Press (1982).
8. C.H. Sauer, and K.M. Chandy, Computer systems performance modelling, IEEE Computer 15:157-158 (1982).
9. R.M. Hord, The Illiac IV: the first supercomputer, IEEE Spectrum 21:15 (1984).

PARALLEL PROCESSING BASED ON ACTIVE-DATA

C. Jesshope

Department of Electronics and Computer Science, The University, Southampton SO9 5NH, U.K.

ABSTRACT

Parallelism as a means to increased computer performance is now widely accepted. Current interfaces to implementations of parallelism in computer systems tend to increase the complexity of the programming task, through elitism, asynchrony and little abstraction away from the underlying implementation. Moreover the most popular abstractions being proposed to limit this complexity, those of the applicative programming school, tend to be based on the process as a means of organizing and controlling the underlying replication. It is the tenet of this paper that the data structure and its composition provide the most efficient abstraction for an implementation of parallelism without pains. Moreover the object oriented paradigm is proposed as a suitable vehicle for its effective implementation.

1. THE EXPLOITATION OF PARALLELISM

The field of parallel processing has made major advances in the last decade; most importantly it is a field of knowledge that is moving from the academic and specialized disciplines into a more widespread industrially oriented field of activity. However this period of rapid growth has provided diversification and not unification, which must be a common phenomena in any active field of research. This paper explores some of the major issues in this field and proposes solutions which to some extent attempt to provide this unification.

There is a wide range of means by which parallelism may be introduced into computer hardware: by horizontal control of multiple function units; pipelining of instructions and data streams; multiple processors; and processor arrays under common control. There is also a multiplicity of programming methodologies to exploit or control this hardware concurrency. At one end of the spectrum we have the large scale scientific users, who have traditionally programmed in FORTRAN, although that language has evolved considerably over its three decades of use. At the other extreme we have the declarative school of programmers, who would like to execute a specification of a problem, rather than an imperative algorithm for its solution. To some extent these two extremes represent another division between applications, that between numeric and symbolic computation. Finally there is a range of control mechanisms which may be used to sequence or activate orders within the computer, which tend to be linked to a programming methodology. For example there is the explicit control found in conventional von Neumann computers, less explicitly there are pattern driven computations, as found in associative processors for example, and finally control is at its least explicit in data flow architectures, where the sequence of instructions executed is determined only by data availability. In data flow control schemes, programs are graphs, where data producers and data consumers are linked by arcs. These arcs can be considered as the memory space of the program and can be implemented as tagged or coloured packets in a multiprocessor system.

There is no doubt that in most general cases, parallelism introduces additional complexity into the programming task. The use of a multiplicity of instruction streams compounds the ability of the programmer to introduce bugs into her programs. There is nothing really new in this, as virtual concurrency has been used for at least a decade now as a means of controlling the complexity of the interactions between the various components of operating system. It is also well-known that operating systems are notorious for their ability to conceal errors, waiting for the appropriate conjunction of circumstances to manifest their ailment.

Deadlock is a prime example of an error introduced through concurrency. Deadlock, or the deadly embrace, is the ability of two or more processes to mutually suspend each other, for example, by each waiting for another to perform an action, which is precluded by the action of another process. The classical example is a ring of processes each trying to read from another, but none willing to write first; were one process willing to write before read, the a write-read-write sequence would ripple around the ring of processes.

Such problems arise because of the asynchronous nature of communication or synchronization between instruction streams in this process based view of concurrency. There are other models however, which avoid the problems associated with this process view. The alternative viewpoint is to consider the data structure and to perform concurrent operations on all elements of the structure. This is the model adopted for obvious reasons by SIMD and vector-based pipelined computers. However it has a much wider applicability than has been exploited to date, as the example below will illustrate.

Because this model has not been well developed, it tends to be dismissed as too restrictive and inefficient. It will be shown here however, that properly developed the model can be applied to a wide range of applications and implemented on many diverse architectures. Indeed, it will be shown that the model developed is far from being restrictive, in that it may efficiently stimulate a MIMD environment on a SIMD machine for example, where the process is replaced by the active-data and load balancing is achieved by the equitable distribution of the active data.

It will also be shown that programming over such a model can be made more attractive and less prone to error than its sequential counterpart. This is because the semantic content of the instruction has been raised without introducing issues such as asynchronisation and deadlock.

2. THE DECLARATIVE VERSUS THE OBJECTIVE STYLE

Even in sequential systems, we have long ago passed to a point where where it was possible to exhaustively verify software. What attempts then have been or are being made to alleviate this "Software Gap"; the difficulty of creating large, complex and, most importantly, correct software systems. One solution has been to introduce more rigour into programming, through functional languages or more recently through logic languages. In both cases a mathematical foundation has been adopted in the development of these languages. In the first we have the use of Lambda calculus, with its tremendous power of abstraction; in the second we have the use of predicate logic, with its equally great power of deduction. In both classes of language, the programs written in them will have a dual reading. The program may be read as a specification, which determines what is computed by the program. Alternatively, the program may be read as a set of rules, which determine how the computation is performed. Because of the first reading, such languages are called declarative.

There are problems associated with the declarative approach however, most importantly one of efficiency. As an efficient declarative program must rely on the programmer favouring the second reading of the program. Such a bias will tend to produce a more imperative style of program, obviating the advantages of this approach's declarative reading. The other disadvantage is the inability of the declarative program to effect its environment. The lack of state and side effect can be unnecessarily restrictive in some applications. I believe that both of these criticisms are fundamental dichotomies introduced by the declarative style. Although it is likely that efficiency may be improved by intelligent program transformation, the declarative style is likely to be limited in its scope of application by the second disadvantage.

Another solution to the software problem, which is more pragmatic than theoretical, is that of the object-orientated programming approach. This is a natural extension of the concepts of modularity and structured design, which found favour in the 1970's. Encapsulation, one of the primary tenents of an object-oriented approach, restricts the access to objects from the outside world to its code or methods, and does not allow access to its data. This changes the sense of action in a system, from one of external manipulation to one of internal or auto-manipulation. Objects can of course interact by participating in the selection of methods from other objects; a message passing environment is usually used to provide this interaction. The second primary tenent of the object-oriented approach is inheritance and its associated class structure. This allows generic objects to provide a mould or factory for the production of user objects, by creating storage and providing access to code or methods. This mechanism also allows the sharing of methods between objects or data types, by the use of a class hierarchy. This can be likened to the abstraction of function over function as found in Lambda calculus. Indeed, given a sufficiently flexible implementation, higher order functions could be defined in this manner.

Object oriented programming provides a solution to the software problem by requiring a more rigorous interface definition between interacting components of a system. The absence of external side effect allows effective modularisation of code. Indeed B.J. Cox (1986) promotes the concept of 'software chips', which could be purchased off the shelf for example. These would come complete with data sheets, to enable the software engineer to construct software systems in much the same way that the electronics engineer constructs hardware systems from off the shelf chips.

The object oriented approach has many advantages as a paradigm for concurrent programming systems. One can for example conceive of communicating processes as objects; certainly Occam already goes a long way towards an encapsulation of its data. However our interest here is one of active-data and data-structure concurrency. It will be shown that the object-oriented approach has major benefits in this area. The goal of this synergism is based as much in implication as in programming style, as it provides a means of encapsulating and abstracting away from the resources in the implementation.

For example an object can be conceived of as:

i) a parallel data structure;
ii) a set of methods applicable to that structure, and
iii) a set of resources on which the methods may be executed.

In the most general case, all three may vary dynamically, as structures, methods and resources may all evolve during a computation. The object must of course maintain an overall invariance, at least within the scope of the common methods offered in an evolving object. An implementation of such a general model will provide a route for binding and hiding different hardware models, such as SIMD and MIMD, and enable the unification referred to in section 1 above. What is more the implementation need not consider the SIMD/MIMD interface directly, as objects implementing the model over each hardware model may themselves be bound by a system or manager object.

3. THE ACTIVE-DATA MODEL OF PARALLELISM

The active-data model of concurrency provides the simultaneous operations across any data structure or object. The operations on the individual elements of the structure take place as if simultaneously, hence even if implemented over MIMD hardware, we have a synchronous model, without deadlock. This model is best illustrated by considering the following example, which contains both an expression of, and an assignment to the same structure or object:

```
A <- e(A,B)
```

In this example the expression over A and B would be evaluated for all elements of A, before an assignment was made back to the structure A. This model is expressed in APL (Iverson, 1962), CmLisp (Hillis, 1975) and the array extensions of FORTRAN 8X.

This model of concurrency can be readily applied to any replication systems, indeed no assumption has yet been made about the control strategy, and implementation of this model could equally well be made over SIMD or MIMD architectures. The major problem however, is one of mapping data onto the available processors and maintaining a high utilization of those resources.

One of the reasons that parallelism has not been exploited as much as it could have been over the last decade, is the issue of mapping the problem domain over the machine domain. To date, this has been the responsibility of the user, who has had little assistance. No model of parallel computation can be considered global unless it describes attributes of a computation, rather than a particular computer system. A counter example is DAP Fortran, which describes computations based on synchronous data operations for a particular array processor. In order to provide portability, and to make this model general, a virtualization scheme must be introduced, which abstracts the model away from the underlying hardware.

The virtualization scheme proposed here (and elsewhere; Hillis, 1985) is based on the data structure, and its activation. The model uses the abstraction of a single virtual processor per data structure element, although at any given time only few virtual processors may be allocated to a real resource (the real processors in the system). This scheme is analogous to a virtual memory for the storage of large structures. Indeed the model may contain a memory virtualization to provide processing over structures too large for the memory requirements, thus virtual processors may themselves reside in backing store.

In practice not all elements of a data structure will require updating, and in a sequentially based language the selection is performed by indexing arrays and by list traversal. Both selection techniques may also be augmented by conditional structures in the programming language. This active-data model is based on concurrently processing the active (selected) data structure elements. Typically the procedure for writing programs in the active-data model can be decomposed into the following sequence:

i) define elements of structure for updating, and
ii) evaluate the method to update the structure.

This partitioning is illustrated schematically in Figure 1, here both stages can be thought of as different methods applied to a data object. In relating this to conventional sequential programming models, the activation would translate into some control structure and the method or code would represent the body of the construct. It is possible to extend this model to one of an encoded activation stage, which could select an appropriate method for each encoding.

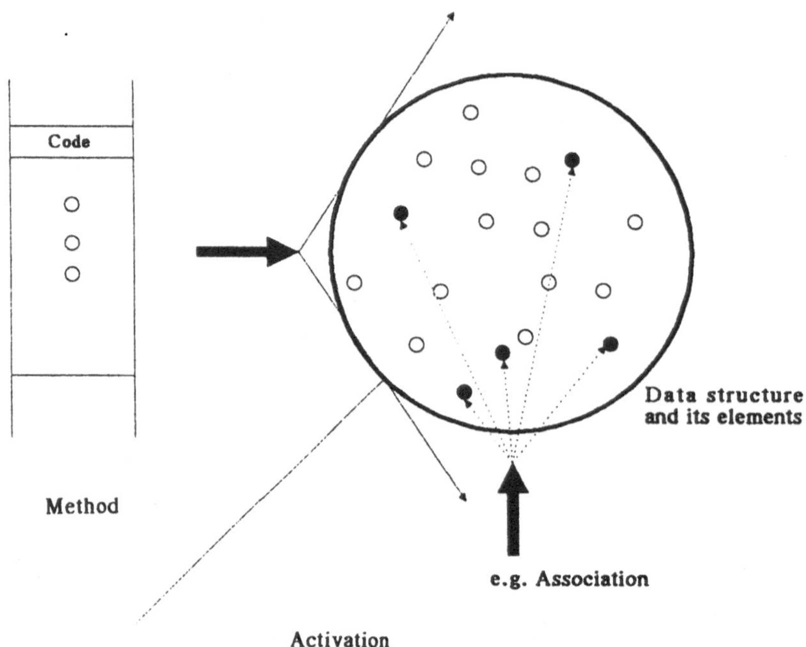

Fig. 1. The partitioning of the active data model into methods and activations.

For example consider an activation similar to that shown in Figure 1, but which coloured a subset of the elements of the structure black and the remainder white. It is possible to use this coding as a method selector. This of course is nothing more than the distributed implementation of the IF ... THEN ... ELSE construct. The important point to note is that the activation may be decoupled from the method and an encoding or tag associated with the data structure elements. Sequencing of operations may thus be data driven, an effective technique for implementing load balancing. For example, provided that both sets more than cover the available resources, then an efficient execution of both methods is possible. Schemes for an equitable distribution of the two sets over the available resources are considered in section 4.

Of course it is possible to bring to bear the whole armoury of structured program design to this methodology, with a hierarchy of data structures and appropriate classes of methods. Figure 1 for example could represent a user selecting a group of files for processing from some graphical interface, in a database application, with the code being built from lower level methods, including further activations, based on data association or discrete function evaluation.

In order to consider the problems of load-balancing, an abstract machine and some implementation details must be considered. The abstract machine used here reflects the duality of the model; the separation of activation and methods. It comprises a set of processors for performing routing operations, to activate and distribute data, and another set of processors to process the data (these would also provide local activation of data).

Figure 2 illustrates the abstract machine. The array of communication processing elements (CPEs) are joined by a communication network and each CPE is also connected, by shared memory, to a processing element (PE), which performs the computation required by the model. Both PE and CPE may require other memory, however a number of data queues are maintained within the shared memory:

Network queue

This is for packet routing only, and provides a queue of packets waiting to be forwarded through the network by the CPE, this queue may be filled by the CPE or the PE and is emptied only by the CPE.

Work queues

A number of queues are maintained to hold active data for currently active methods. These hold data packets waiting to be processed by the PE, these queues may be filled by the CPE or the PE, but are only emptied by the PE.

The network may be of any topology, but simple routing strategies exist for regular networks. Implementation restrictions

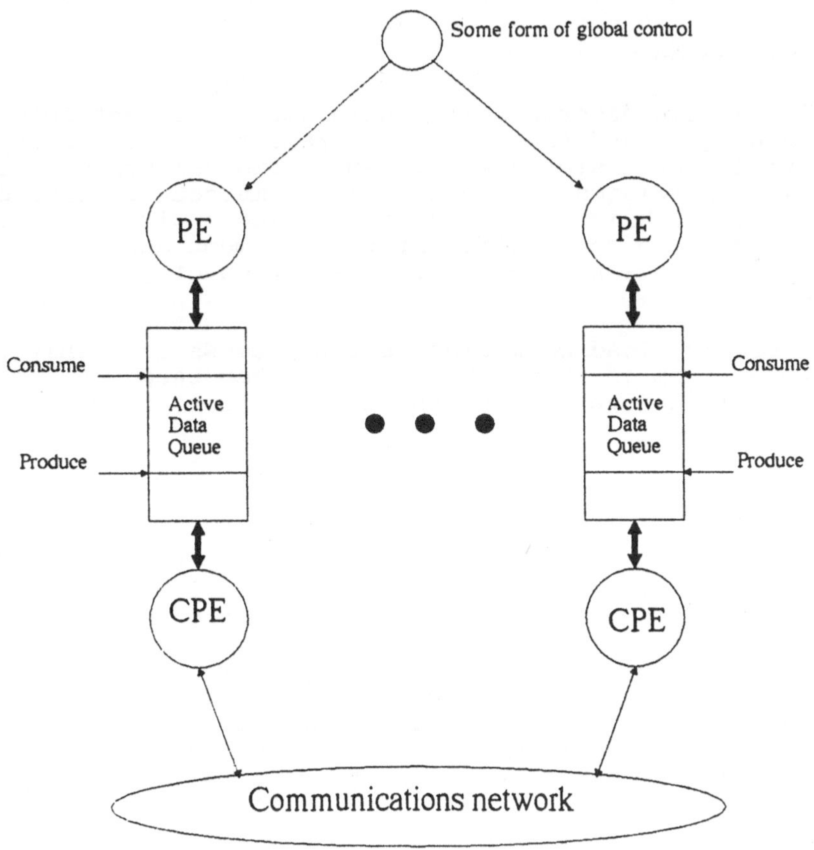

Fig. 2. The active data abstract machine.

are likely to define its topology. Ideally on each routing cycle, the CPE should receive and process one data packet from each direction in the network. The processing required simple differentiates which queue the data should be placed in, the network queue, for further forwarding, or one of the work queues if the packet has arrived at its destination. The destination may be defined by an address held within the packet, or by some state held locally in the CPE. This latter situation occurs in some load-balancing algorithms. Within the same cycle, the CPE should also forward one or more packets from the network queue.

A second network, the global control, links the PEs in a fan in/fan out tree. This is to provide broadcast and reduction operations, which are central to the active-data model. The requirement for those control mechanisms is illustrated by the simple example of operations over sets, where the most frequent operations involve both of these global communications operations. The set is one of the most powerful data structures in many symbolic applications. It is used extensively in load-balancing algorithms. Examples of set operations are:

is_in : requires association of an object with every member of the set, and

max : requires the reduction over the st using the greater than operation.

4. LOAD BALANCING

This section describes and enumerates different load balancing techniques and flags the requirements for efficient implementation. Both static and dynamic load balancing will be considered and related to the activation methods described above. These techniques have direct analogue with process load balancing, but are more efficient to implement and moreover are more closely representative of the semantics of parallel algorithm development.

Put simply, load balancing is the process of evenly distributing entries in the work queues, either based on local or global information concerning the distribution of work.

The techniques can be classified as follows:

a) Static deterministic

We can assume that for static structure, an even distribution of data structure elements over the available resources will be made at compile time. During execution however, substructures may be activated from the evenly distributed source structure. In the worst case for example, it is possible that all activated data may be associated with a single processor. A more realistic situation is illustrated in Figure 3, where a row has been selected from a matrix, mapped over the available resources. The two axes represent processors and memory, or alternatively parallel and sequential execution.

It is possible with a statically distributed regular structure to redistribute the elements of a selected sub-struc-

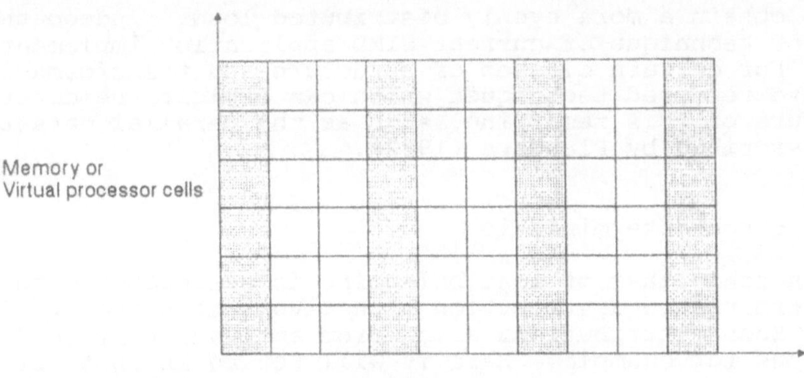

Fig. 3. Static deterministic activation, from substructure selection for example, such as a row or column from a matrix.

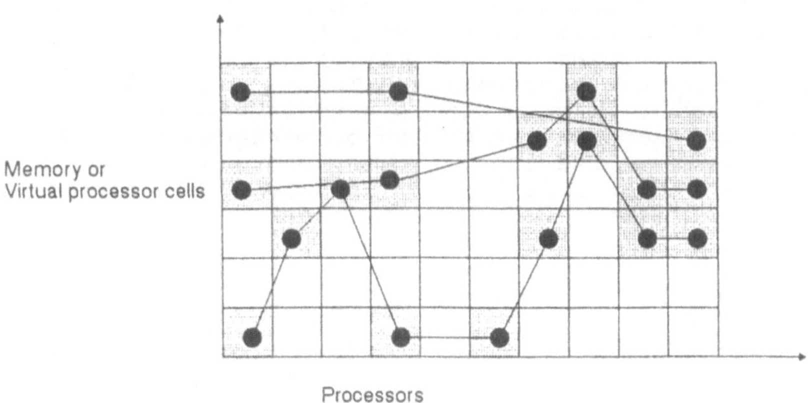

Fig. 4. Static nondeterministic activation, by association or local conditional evaluation. Diagram shows layers of concurrent operation, that arise from the use of a queue-of-data implementation. Note that there is no migration of data.

ture to obtain a more evenly distributed load. Indeed this is the major technique of current SIMD application implementations. For certain classes of structure and transformation, there are reported techniques which can exploit the deterministic nature of this remapping, such as the parallel data transforms described by Flanders (1982).

b) Static non-deterministic

A second class of load balancing is associated with the non-deterministic distribution of active data structure elements. Such distributions arise from activation by conditional operations for example. Here it will not be known a priori, where the activated elements will be found. Provided that it can be expected or determined that a reasonably even distribution of active elements will result, the use of work queues as defined in the abstract machine, will allow automatic load balancing in such circumstances. This is illustrated in Figure 4, where it can be seen that given active element queues, only two passes of the of the method would be required instead of 4. In current SIMD implementations, such as the ICL DAP, this technique is difficult and expensive to implement, as the distributed queue structure requires local addressing, which the DAP does not support directly. It can of course be simulated over a memory block in logarithmic time, by redistributing blocks, using the bit address as an activation mechanism. There is a trade-off therefore in the complexity of the method implemented, for a simple method the technique would not be appropriate.

c) Dynamic non-deterministic

The final class of load balancing is the most general and would provide load balancing in all eventualities. The situation in which this would be used is where an uneven and non-deterministic distribution of active data structure elements would result. For example this may occur in discrete function evaluation or marker propagation. The situation is illustrated in Figure 5. What is required is a redistribution of active data structure elements, which can be achieved by local or global reference. In a local scheme, packets of data would be distributed based on local comparisons over work queue lengths. However this is likely to result in a great deal of unrequired activity in the communications method.

A global scheme may be implemented by taking a histogram of the distribution of the load, from which a desired average load could be established. From this desired load, a routing strategy can be adopted, which forwards packets of information, not on address, but on local load compared to desired average load. For example, if the average load is broadcast to all processors, through the global control network, then if the processor's local load is to within some bounds greater than the average load, then it would emit packets; if it were less than the local load, it would accept packets; otherwise it would simply forward packets. Direction of forwarding would need to producing a randomizing effect, but also adapt to traffic density.

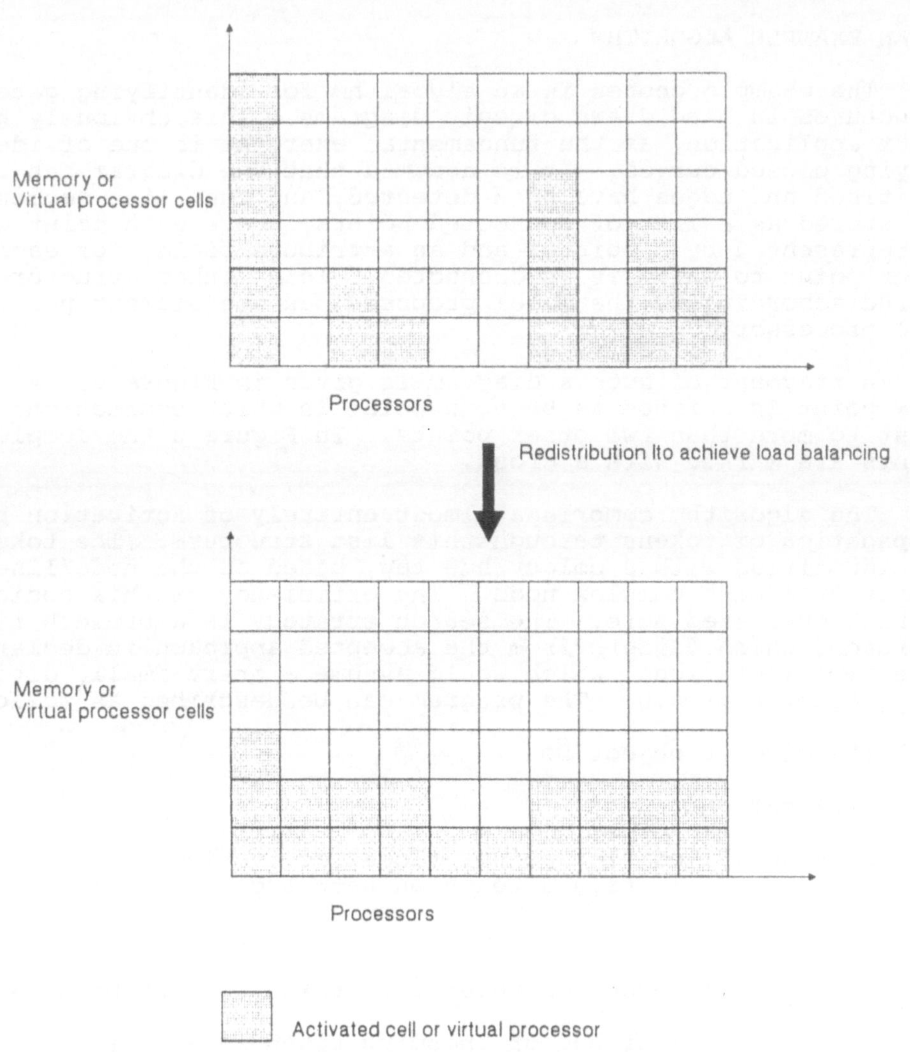

Fig. 5. Dynamic nondeterministic activation, for example by token propagation through the communication network. Queues are redistributed (virtual to real processor distribution is dynamically modified according to load).

The problem in load balancing linked structures is in maintaining the integrity of pointers across the array. The migration of virtual processors can of course be achieved by addressing virtual nodes, rather than physical nodes and having a routing strategy which adapts to any load balancing migration. Alternative data may be diffused temporarily to achieve a balanced load for processing a particular method, and then by attaching return addresses or 'springs' to the data packets when farmed out, the original connected structure can be recreated, by an addressed-packed, final-routing phase.

5. AN EXAMPLE ALGORITHM

The example chosen is an algorithm for identifying gate structures in hand drawn circuit diagrams. This obviously has wider application, as the fundamental exercise is one of identifying closed curves. It is assumed that the diagram has been digitised and edges have been detected, and that line segments are stored as a list of connected points, where each point will be represented by a pointer and an attribute field, for each other point to which it is connected. This linked structure is stored according to the model proposed, as one element per virtual processor.

A fragment of such a diagram is given in Figure 6. A complex point is defined as being a point in which connections exist to more than two other points. In Figure 6 the complex points are marked with a cross.

The algorithm comprises almost entirely of activation by propagation of tokens through this list structure. The tokens are identified with a colour/hue tag, based on the node/line number from each complex node. The efficiency of this coding is not considered here. The search strategy is a breadth first approach, which differs from the accepted approach in declarative implementations, which would assume a start-small, divide and conquer approach. The program can be described as follows.

With the circuit object Do

* Initialisation

 i) Activate all complex points (local activation),
ii) emit a uniquely tagged token on each arc.

Do

 i) Activate nodes on receipt of token (marker propagation).
ii) Store colour/hue of incoming token in a "I saw" structure;
    i) Activate complex nodes

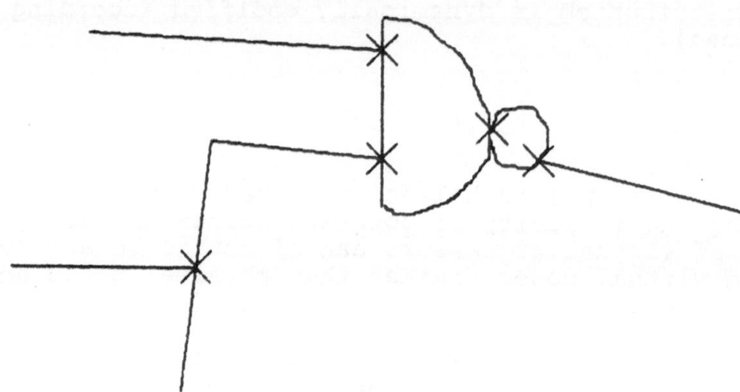

Fig. 6. A fragment of a hand drawn schematic, showing complex points.

ii) Modify local colour/hue with own colour/hue, at complex points only;

emit a token to all points, except the point from which a token was received.

Until termination (No more active elements)

In the algorithm described above, the phases i) and ii) represent the activation and application of method to active data respectively, which are nested in the repeat loop. It should also be noted that the algorithm as it stands is very inefficient, indeed it will never terminate. We must therefore consider token removal and optimizations.

Token removal: Tokens may be removed when:

a) a singleton point is reached, or
b) a packet reaches its sender, i.e. a packet is received, which contains the local colour as its base colour. These tokens must be stored in a local success structure.

Closed path identification: A closed path may be identified by broadcasting a colour/hue combination. This must be associated within the "I saw" structure, to light up the corresponding closed path.

Optimizations: There are optimizations that may be made to this algorithm, but not all may be made safely, without a priori knowledge of the nature of the diagram. The first optimization, which may be safely made, is in omitting one of the lines in the initialisation of the algorithm. If that line is not a part of a shape, this does not matter. If it is a part of a closed shape, then a packet will have been emitted in the other direction around the Eulerean path.

A second optimization, which may also be safely made, is to kill all sibling tokens once a closed path has been established for a given colour/hue combination.

A third optimization, which may not always be safely made, is to remove all lines from contributing to the algorithm after they have been identified with a closed path. This works for simple closed paths, but not those where line segments are shared between more than one closed path.

A final optimization would be to terminate packets on some other criteria, such as length of lines, where for example it may shown that a line of a given length could not possibly contribute to a gate. These are heuristic optimizations, which may vary from application to application. However, it should be noted that using the object oriented approach to software development, the safe optimizations may be built into the general method and optimizations may be added by the user for her particular application.

6. THE IMPLEMENTATION ISSUE

Little reference has been made to implementation issues concerning the abstract machine. There are, however, a number of issues that must be taken into consideration, when choosing an implementation.

First consider the MIMD/SIMD structure. Whereas a MIMD implementation may run different methods on different processors, and this implementation would be ideal for a data driven object oriented paradigm, where it can be conceived that the packets flowing within the system may contain method selectors, it has already been shown that by exploiting load balancing techniques, multiple methods may also be efficiently implemented on SIMD implementations. The methods are executed in sequence and provided that there are sufficient active elements for each method, then a full utilization of the available resources may be achieved. There are also positive advantages for SIMD implementations; in many methods there is a requirement for broadcast and reduction communications within the PE structure, see for example the algorithm in section 7. In a SIMD machine, these operations would be implemented as part of the global control scheme. Data within the global control word may be used as an associative key, and a reduction by logical sum tree of a value in each PE is a common SIMD control mechanism; the root of the tree being sensed by the global controller. A MIMD implementation would have to source the broadcast data from either a nominated processor or could perhaps source it from any processor. Moreover the realization of broadcast and reduction is likely to be provided by the slow action of distributed communicating processes, rather than by electronic (or optical) signals, as could be implemented on a SIMD machine.

At Southampton we have been investigating the implementation of this model of parallelism over a SIMD structure. An exploratory implementation of a packet routing communications scheme has been implemented on the RPA computer system, which currently exists as a simulator only (Jesshope and Stewart, 1986). The RPA processing element (PE), like most other SIMD PEs selects an input from one of a number of neighbouring PEs, in this case from one of the four orthogonal neighbours. However, unlike most other arrays, the RPA PE is able to locally choose the direction selected. This local asymmetric behaviour is of great benefit in mapping regular data structure over the RPA array, and is described in more detail in Jesshope et al. (1987) and Jesshope (1987). It will be shown here that this also provides for sufficient utilization of the communications structure, when implementing the packet routing system.

In order to better understand the implementation, a brief description of the PE is required. The RPA PE has a 2-bit source and 2-bit result bus. The result bus can be connected, using a local control field, to the source bus of one of the four neighbours. Two bits of data may be passed into and out of each PE in a single cycle, over the same leg if necessary (i.e. north selects south and south selects north). The storage in each PE comprises two eight bit stacks (capable of pushing two bits per cycle) and 64 bits of RAM organized as 8x8 block, with byte-wide, parallel-to-serial conversion provided by two shift registers between the sources and destination busses. These shift registers also contain parallel comparitor circuits (giving <, >, and = as two bits enabled onto the source bus). It is also possible to locally address the word store.

In a packet switched communications scheme, the address of the data is provided at the source, as a part of the packet, but in the RPA hardware, data must be selected at its destination. We therefore have to implement a protocol to invert the sense of the direction control from a 'send to' to a 'receive from' address, and resolve any contention that may result. This protocol is implemented in microcode and must establish as many channels between PEs as possible. This is achieved by a sequence of polling operations. For obvious reasons we have chosen a local but deterministic algorithm, which is efficient to implement. Although it does not provide the optimal solution, it provides a good compromise between the number of channels implemented and the effective bandwidth over those channels.

Absolute data packet addressing is used, because the PE can provide rapid routing information using the eight bit comparitor. Although we anticipate building a 32x32 RPA, this scheme would support arrays of up to 128x128.

Protocol: In the following, a set is mapped over an array of PEs with one member of the power set mapped onto each PE. To implement the protocol we define the following signals:

R_i $i \in \{n,s,e,w\}$, the set of PEs requiring channels in n,s,e and w respectively. $\exists\, R_i \cap R_j =$ <empty>,

A_i $i \in \{n,s,e,w\}$, the set of PEs having their requests acknowledged. $\forall\, i,j\;\; R_i \cap R_j =$ <empty>,

Si $i \in \{n,s,e,w\}$, the set of PEs enabled to receive from a given direction, and

W set of PEs currently able to accept a channel.

A function shift(X,dir) is required to describe the protocol, which shift the set X of signals in the direction dir {n,s,e,w}. The algorithm to establish a set of channels is given in Figure 7 below for one priority step. The use of four

```
North channel:      An <- Rn
                    Ss <- Shift (An,n)
                    W  <- Ss
                    Re <- Re-An
                    Rw <- Rw-An

East channel:       Ae <- Shift(W,w) ∧ Re
                    Sw <- Shift(Ae,e)
                    W  <- W-Sw
                    Rs <- Rs-Ae

South channel:      As <- Shift(W,n) ∧ Rs
                    Ss <- Shift(As,s)
                    W  <- W-Sn
                    Rw <- Rw-As

West channel:       Aw <- Shift(W,e) ∧ Rw
                    Se <- Shift(Aw,w)
```

Fig. 7.

steps provides an even priority on all directions, by rotating the priority directions between steps.

This scheme has been implemented and requires between 14 and 22 microseconds to forward up to 1024 32 bit packets through the array, assuming a 100 nanosecond cycle. Larger packets will require a smaller overhead and will asymptote to 150 nanoseconds per bit. For the 32 bit packets, this represents a total possible bandwidth in a 1024 PE RPA of $2x10^9$ bits per second.

7. VLSI: THE FUTURE AND THE SEQUENTIAL MOULD

Replication is likely to be the major mechanism for increasing computer performance in the coming decade. This trend is already been observed in commercial computer designs, in the class of computers known as near supercomputers, such as AMT DAP, Meiko computing surface, Alliant FX/Series, Intel IPSC, NCube and Sequent Balance, all of which contain multiple processors. The reason for this surge in interest in this general class of architecture are the needs of VLSI, which are twofold; economic VLSI designs require:

i) regular layout with regular interconnection patterns, and

ii) the economics of scale.

Both of the properties above are found in memory chips and this contributes directly to their low cost. Unfortunately memory does not contribute o increased processing performance. Indeed it perpetuates what I call the sequential mould of programming, which actively discriminates against the successful exploitation of replication.

The sequential mould states that: "a machine may have unused resources, providing that these resources are memory, and NOT processors". Nobody minds running a 5 Mbyte application on a 10 Mbyte machine, but to use only 500 out of 1000 available processors could be considered as a heinous crime. This paper has presented an abstract model of concurrency, which attempts to alleviate the insecurity felt by programmers in the sequential mould, when not all processors in the system are active. It does this by providing dynamic load-balancing over a distributed data model of concurrency.

Of course the introduction of such a scheme necessarily introduces overheads, which must use a finite amount of the machine's resources. The overheads in this model are in general small, compared with the analogous model of dynamic process-based load balancing. Indeed this is the key thesis of this paper; the data abstraction of processing is inherently more suitable for efficient utilization of highly replicated computer resources than the process based abstraction.

8. SELECTIVE BIBLIOGRAPHY

Cox, B.J., 1986, "Object Oriented Programming", Addison-Wesley.

Flanders, P.M., 1982, A unified approach to a class of data movements on array processors, IEEE Trans. Comput. C-31:405-408.

Iverson, K.E., 1962, "A Programming Language", Wiley.

Jesshope, C.R., Rushton, A., Cruz, A., and Stewart, J., 1987, The structure and application of RPA: a highly parallel adaptive architecture, in: "Highly Parallel Computers", Elsevier Science Publishers, North-Holland, p. 81-95.

Jesshope, C.R., 1987, The RPA as an intelligent transputer memory system, in: "Systolic Arrays", Moore, McCabe and Urquhart, eds., Adam Hilger, p. 283-293.

Jesshope, C.R. and Stewart, J.M., 1986, MIPSE - a microcode development environment for the RPA computer system, in: "Software Engineering 86", Barnes and Brown, eds., Peter Peregrinus, p. 184-196.

ARCHITECTURES FOR SIMULATION ENVIRONMENT

G.C. Vansteenkiste and E.J.H. Kerckhoffs
University of Ghent
Coupure Links 653, B-9000 GENT, Belgium

ABSTRACT

The utilization of some parallel and pipeline computers in continuous systems simulation is the topic of this paper. Particularly, the attention is focused on the MIMD structured Delft Parallel Processor (DPP81) and the Applied Dynamics AD10 system, and their use in solving simulation problems. Additional to discrete time parallel processing, also continuous time parallel processing is briefly reviewed. Here, the emphasis is on the EAI SIMSTAR system, which recently appeared on the market.

1. INTRODUCTION

In parallel processing the computing capacity is distributed over a number of processing elements, which are able to operate in parallel. In this paper by "distributed simulation" is meant simulation on this type of nonconventional computers. The oldest form of parallel processing is analog computation, which is based upon continuous time processing. Continuous time parallel processing is briefly discussed in Section 2.

Signal processing and large-scale simulation, that both ask for number crunching facilities, have stimulated to a large extent the development and utilization of digital computer systems with architectures substantially different from those of the conventional von Neumann computers. Especially worthwhile in this respect is the introduction of the multiprocessing and pipelining concepts in some new computer architectures (Hockney and Jesshope, 1981; Spriet and Vansteenkiste, 1982).

In multiprocessing an important approach is the attainment of parallelism through the replication of one or more of the major functional units (ALU, memory unit, control unit) in some computers, in order to increase the computational speed and power. In SIMD (single instruction stream, multiple data stream) computers a number of processing elements, each consisting of an arithmetic/logic unit and a memory unit, are interconnected to form an array of processors under control of a single control unit (Flynn, 1972). SIMD computing systems were designed in the 1960's and early 1970's to achieve high computing speed, particularly in the treatment of partial differential equations by finite difference methods, without the need to replicate the relatively expensive control unit. The classic example of such SIMD computers was the (now dismantled) ILLIAC IV (Barnes et al., 1968; Kuck, 1968). With the advent of more advanced hardware technology and decreasing cost of digital hardware, replication of the control unit became feasible; as a result, a variety of computer systems have been constructed containing an array of processing elements, where each element now includes an arithmetic/logic unit as well as a control unit. Unlike the SIMD systems, the different processing elements carry out different logic and arithmetic operations. For this reason, systems of this type are referred to as MIMD (multiple instruction stream, multiple data stream) computers (Flynn, 1972). Except for the commercially available HEP manufactured by Denelcor Inc. (Brinton, 1982), most MIMD computers are developed, built and operational within university or other research environments.

In section 3 the emphasis is on one such MIMD computer: the Delft Parallel Processor DPP81.

Pipelining techniques are particularly effective in enhancing execution speeds in the case of processing loops. Pipelining is applied in most so-called supercomputers (such as CRAY-1 and CYBER 205), developed for very effective vector processing (Johnson, 1978). A basic aim in constructing algorithms for vector processing is to perform the bulk of the computational tasks on one-dimensional arrays, i.e. vectors. Notwithstanding the value of vector operations in simulating continuous fields, many important large-scale problems cannot be organized into vector form efficiently (e.g. problems that call for much searching and sorting, or that are dominated by conditional branching). Pipelining techniques are also employed in the so-called peripheral array processors (Louie, 1981). Although nearly all array processors are signal processing oriented (such as e.g. the AP120B manufactured by Floating Point Systems), they are also extensively used for simulation purposes (Karplus, 1977). However, from the simulation point of view array processors have the disadvantage that they are especially designed to perform relatively compact operations on large blocks of data, while in simulation normally during each frame time lengthy and elaborate computations must be executed on small sequences of numbers; moreover, no facilities are available for direct access to the processor from external communication lines. An array processor, especially developed for real time simulation, is the AD10 manufactured by Applied Dynamics International, which is discussed in Section 4.

2. CONTINUOUS TIME PARALLEL SIMULATION

2.1. Continuous time processing

In continuous time parallel processing the arithmetic operations are performed in a continuous time set T_c

$$T_C = \{t \mid t \in [t_o, t_e]\}$$

This implies that each computing component needs input and generates output at any t of time set T_C; since also the information exchange takes place in the same time set T_C, no synchronization problems arise when connecting the various components. Another feature inherent to continuous time data processing is the natural performance of "interval mappings" (Dekker, 1976) of both the instantaneous type:

$$y(t) = f(x(t)), \forall t \in T_C$$

and non-instantaneous type:

$$y(t) = \sigma(y_o, x[t_o,t], t-t_o), \forall t \in T_C \quad (1)$$

In (1) 0 is a mathematical operator and x(t) a (piecewise) continuous signal. A practical example of the latter is the "integration with respect to time t", which means that - unlike discrete time processing - continuous time processing has integration as a basic arithmetic operation. Finally, in continuous time processing dependent (!) tasks can be executed in parallel. Taking for instance the "interval mapping" $y(t) = x1(t)x2(t) + x3(t), \forall t \in T_C = [t_o, t_e]$, in discrete time processing this mapping can only be performed a finite number of t-values on the interval Tc; for any t the multiplication and addition are dependent tasks that cannot be done simultaneously. In continuous time processing they can be done in parallel, since "successive approximations" are actually calculated and transferred between the various computing elements. The capability of performing in parallel both dependent and independent operations causes continuous time parallel processing to be true parallel processing, i.e. a complete computing time invariance is achievable with respect to the problem size and complexity.

The practical realization of continuous time parallel processing is in analog computation, which actually is the oldest form of parallel processing (Korn and Korn, 1964). A hybrid computer is a combination of a digital and analog computer with the necessary interface equipment; in fact, the analog part is a parallel operating peripheral of the digital computer (Bekey and Karplus, 1968; Feilmeier, 1974). The so-called "autopatched" hybrid computer has programmable switch matrices to interconnect the various analog and logic components (Brok et al., 1978; Brok et al., 1979). A time shared hybrid computer is an autopatched computer with additional provisions to multiplex the analog part over several processes. The practical advantages of the hybrid computer are its computing speed, its ability to solve differential equations and its interactive features. The disadvantages are concentrated around scaling, reproducibility, accuracy and the available set of continuous time arithmetic operations.

2.2. The EAI SIMSTAR System

Further evolutions in the continuous time parallel processing field resulted in the EAI SIMSTAR system. The SIMSTAR is an attached multiprocessor system, announced by Electronics Associates, Inc. as the fastest simulation system on the market today (SIMSTAR Technical Description, 1983; Landauer, 1983; Embley, 1984; Ilid, 1984). The SIMSTAR system architecture includes two Parallel Simulation Processors, each consisting of a Parallel Logic Unit (PLU) for Boolean function generation and a Parallel Mathematical Unit (PMU) with associated monitoring devices (see Figure 1). The PMU is the analog subsystem to do continuous signal/data processing and is composed of among others Mathematical Computing Blocks interconnected via a three stages Block Connection Matrix. This Block Connection Matrix is a solid-state switch array which allows any Mathematical Computing Block output to be connected to any Mathematical Computing Block input. The Mathematical Computing Block combines the basic computing units such as integrators or multipliers or summers together with assigned coefficient units, constant units, D/A-switch units, feedback gain units, voltage limiter units and all the digital control circuitry required to allow complete programming and readout from a digital computer.

The SIMSTAR incorporates two main digital computers, that the user must be concerned with: the Digital Arithmetic Processor and the Host Data Processor. The latter is utilized for processing that is non real-time (e.g. off-line program

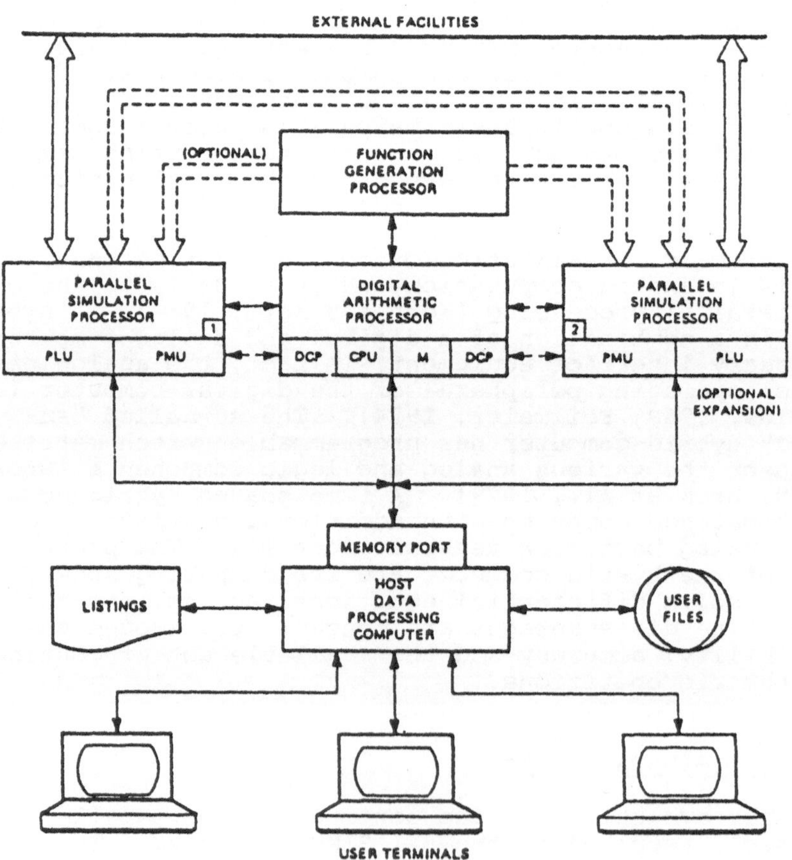

Fig. 1. The SIMSTAR system architecture.

preparation, ACSL all-digital runs, STARTRAN translation, FORTRAN and parallel translation) or does not directly involve the Parallel Simulation Processor hardware (e.g. post-run processing, such as data analysis/reduction, plotting and printing. In fact the Host Data Processor (which does not even run under a real time operating system) is a standard, multiple user, digital system operating in parallel with the attached SIMSTAR Simulation Processor. The Digital Arithmetic Processor performs the remaining tasks: simulation set-up and check-out (initial region), real-time simulation functions (dynamic region) and post-run processing (terminal region). This partitioning of the total computational job allows the dedicated Digital Arithmetic Processor software system to be optimized and controlled separately from the Host Data Processor environment.

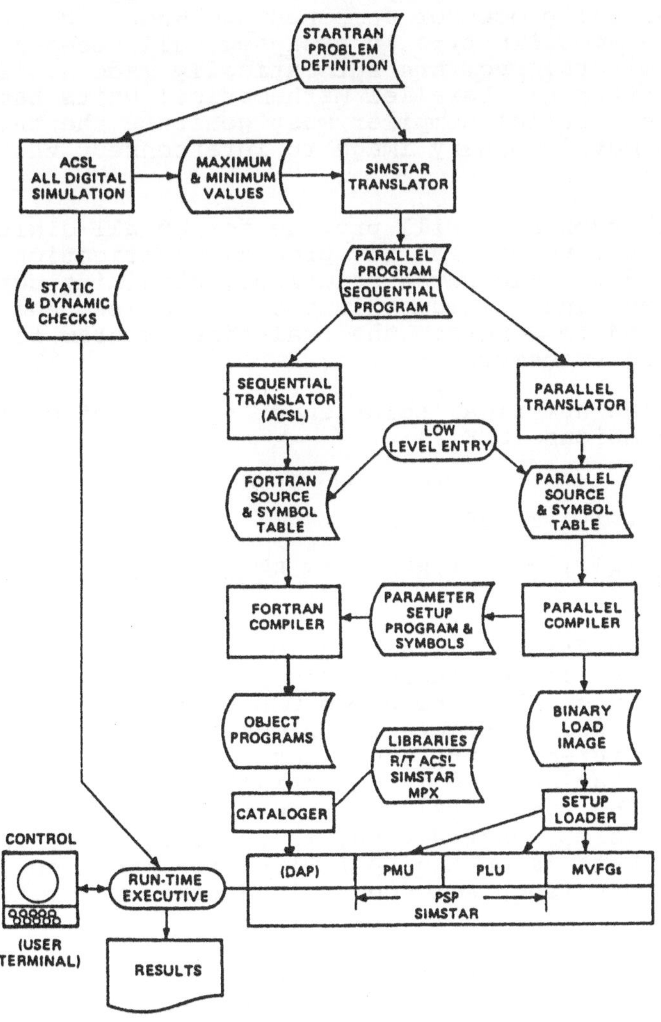

Fig. 2. SIMSTAR Program Generation System flow diagram.

The SIMSTAR software is developed to best implement such attributes as transparency, maintainability, flexibility and user friendliness. In the following, the attention is only focused on the so-called Program Generation System (see Figure 2). This software system (that runs in the Host Data Processor) provides the user with the capability to program the four dissimilar processors in the SIMSTAR from a single source program (compatible with an all digital approach).

There are three main levels of processing:

- the SIMSTAR translator outputs the program split into its parallel and sequential parts
- the sequential portion passes through the ACSL Sequential Translator which provides a FORTRAN source program to be then compiled and linked
- the parallel part passes through a Parallel Translator which produces a FORTRAN-like code to be then compiled into a binary object program for the various parallel processing units of the SIMSTAR. In generating the binary code for the Parallel Simulation Processors the first step that must be performed by the parallel compiler is the assignment of operations to actual processor components. When a computing component of a specific type is assigned, all necessary coefficients, limiters, etc. are automatically made available along with it. After all Parallel Mathematical Units have been assigned, the Parallel Compiler must generate the three stages Connection Matrix binary image to interconnect the components.

The ACSL subsystem will provide for an all-digital simulation capability, to be used for program verification, generating static and dynamic check solutions, obtaining scaling data via the minimum and maximum values of variables, and additionally it is used to generate the real-time program for the Digital Arithmetic Processor.

Unlike the numerical solution of systems of ordinary differential equations (ODE's):

$$\frac{dy}{dt} = f(y,u) \; ; \; y(t_o) = y_o$$
$$t \in [t_o,t_e] \; ; \; y, f \in R^n \; ; \; u \in R^m \; ; \; m \leq n \qquad (2)$$

where one is confronted with a lot of different numerical integration methods, in continuous time parallel processing there is only one basic method to solve ODE's. This method is globally characterized in Figure 3. Block 1 contains n continuous time (analog) integrators, and block 2 n analog function generators. The solution time is essentially independent of the dimension n of the system (2). The speed advantage of the analog solution is the more attractive, if the frequently repetitive solution of ODE's is on the spot with, for instance, varying parameter values (such as appears in dynamic parameter optimalization).

Inherent to the analog solution of Eq. (2) there are two main problems simulationists are confronted with. One kind of problems arises, if the needed functions of dependent variables and/or of the independent variable t cannot be directly gener-

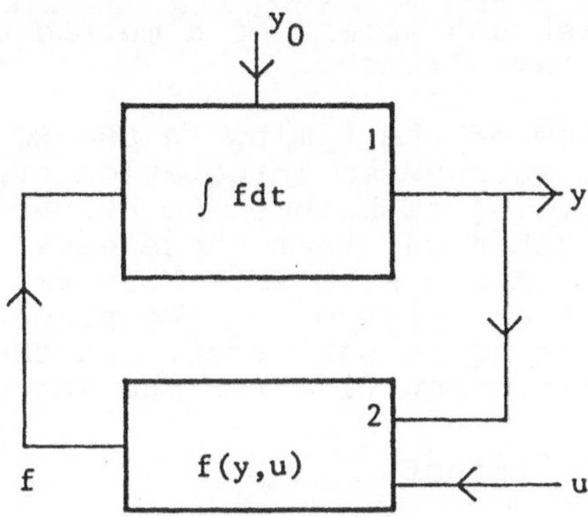

Fig. 3. Principle of the analog solution of ODE's.

ated in the analog part. Piecewise linearization techniques and spline approximation techniques can then be applied to overcome this drawback (Dekker and Kerckhoffs, 1974). Another kind of problem is concerned with the impossibility to solve a large system of ODE's directly on the analog computer because of its insufficient size in terms of available computing components. (This can happen, for instance, if semidiscretization methods are used to solve field problems; in these so-called methods of lines all dependent variables but one are discretized, and a given PDE converts into a (large) system of coupled ODE'S). These troubles can be reduced by parallel overlay and block iteration techniques (Kerckhoffs et al., 1973; Kerckhoffs, 1982).

3. MIMD PARALLEL COMPUTERS IN SIMULATION

3.1. Parallel Implementation of Well-Known Integration Algorithms

Suppose, for the solution of system (2) we use an explicit multistep numerical integration method:

$$y_{j,i+1} = g(y_{j,i}, y_{j,i-1}, \ldots, y_{j,i-p}, f_{j,i}, \ldots f_{j,i-q}, h)$$

$$j = 1,\ldots,n \; ; \; i = 0,1,\ldots \qquad (3)$$

$$f_{j,i} = f_j(\underline{y}_i, u_i) \text{ with } \underline{y}_i = (y_{1,i}, \ldots, y_{n,i})^T$$

Having available an MIMD structured array of m processing elements (PE's), a straightforward approach to solve the set of equations (3) is to partition the set and then to allocate a certain number of the n equations to each of the available PE's. Each PE is responsible for performing the function evaluations and integrations associated with its assigned equations. These arithmetic operations can occur in parallel, with the y-values necessary to do succeeding function evaluations being communicated periodically between the PE's. This method is referred to by Franklin (1978) as the equations segmentation (ES) method; Yura (Yura et al., 1975) considers all function evaluations and integrations assigned to a certain PE as a large (compound) task and speaks of a method of "parallel processing on a large task level".

A pictorial view of the timing in the ES method is given in Figure 4 for two successive integration steps; here, the index set $J = (j = 1,...n)$ is assumed to be subdivided in m subsets $J_1,...,J_m$ in such a way that those equations, for which $J \in J_k$, are implemented in the k-th PE. Moreover, no interprocessor communications are supposed to take place when computational activities are being performed. On the basis of this diagram the total execution time for one integration step is:

$$TES = t_{fymax} + t_{transf}$$

where $t_{fymax} = \max(t_{fy1},...,t_{fym})$ - t_{fyk} being the time to perform the function evaluations and integrations assigned to the k-th PE - and transf is the time needed to effect the information transfers. It is convenient to also denote how the computational work is balanced between the m PE's. This may be done by defining the measure of dispersion

$$D_{fy} = \frac{m t_{fymax} - \sum_{i=1}^{m} t_{fyi}}{\sum_{i=1}^{m} t_{fyi}}$$

It is easy to see that D_{fy} fluctuates from 0 (if the operation times of each of the m PE's are equal) to m-1 (if all equations are placed on one PE). This measure can be used to indicate how equitable the processor scheduling algorithm is in allocating equations to PE's. Note that although the equations can perhaps be distributed over the PE's in a more evenhanded fashion, thus decreasing D_{fy} this could well result in an increase in the communications time.

Since function evaluations and integrations are done in parallel, a good deal of time can possibly be saved over an

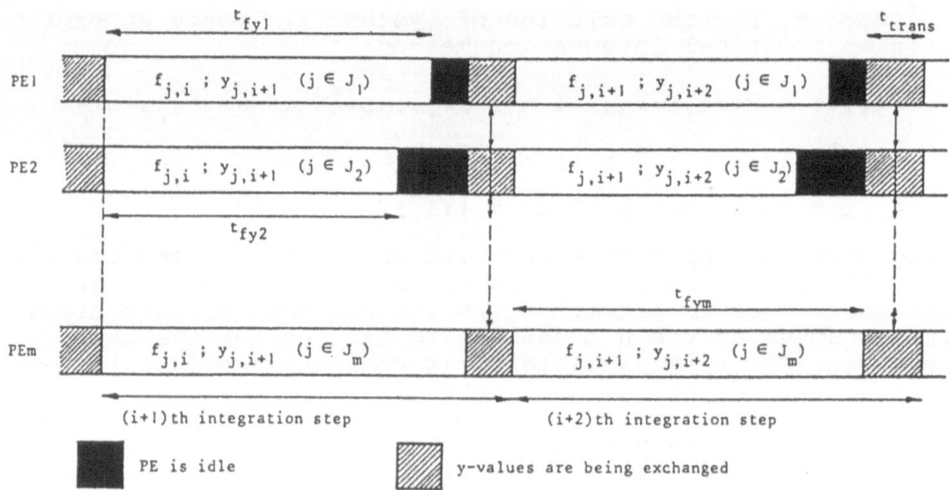

Fig. 4. Timing for ES method (It is assumed that $t_{fym} = \max(t_{fy1},...,t_{fym})$).

uniprocessor procedure. How much time is saved depends on the multiprocessor system concerned (especially the intercommunications characteristics, affecting the amount of communications delay at each integration step), the chosen integration algorithm, the interrelation structure in the system of equations to be solved and the way the equations are allocated to the PE's (which also effects the amount of information transfers).

The above-described ES procedure can be augmented with a partitioning of the function evaluation themselves: (compound) parts of the function evaluations are distributed over the PE's, involved in the ES algorithm (or eventually additional PE's) in such a way that the PE work load is to some extent equally spread. In this manner the affecting computing time for one integration step, actually determined by the PE that performs the most time-consuming (sub)function evaluations, can be decreased; however, additional data exchanges between the PE's are generally required.

If the evaluation of the right hand member of Eq. (3) - i.e. the function evaluations plus integrations - is partitioned on a basic operator level, we have what Yura (Yura et al., 1975) called a method of "parallel processing on a small task level". In general, additional PE's are required to take advantage of the operator parallelism in practice. The existing results from the theory of parallel algorithms (especially those with respect to arithmetic expressions and recurrent equations (Hossfeld, 1983) might be relevant, when solving ODE's with the exploitation of operator parallelism.

In the above, parallel implementations of well explored serially-oriented integration methods have been dealt with (i.e. the partitioning is across the system of equations and the function evaluations). Methods, which are parallel in nature (i.e. the partitioning is across the algorithms concerned), are beyond the scope of this paper: the reader is referred to the literature (Nievergelt, 1984; Miranker and Liniger, 1967; Yura et al., 1975; Worland, 1976,; Franklin, 1978; Yen and Cook, 1982).

3.2. The DPP81 and its use in simulation

At Delft University of Technology (The Netherlands) the MIMD-structured Delft Parallel Processor DPP81 has been built, which has been operational since 1981 (see Appendix).

In continuous time simulation the parallel implementation of numerical integration methods is obviously important. In the ES method, as outlined in section 3.1, the partitioning is across the equations and - provided the number of available PE's is sufficient - it can be continued so far that each equation is solved in a separate PE. As an illustration of this, in Figure 5 the framework of a parallel program is shown, meant to run in the i-th PE, if the DPP81 implementation is considered of the fourth order predictor-corrector (Adams-Moulton pair) integration algorithm:

$$\begin{aligned} Y^p_{n+1} &= Y^c_n + \frac{h}{24}\left(55\,F^c_n - 59\,F^c_{n-1} + 37\,F^c_{n-2} - 9\,F^c_{n-3}\right) \\ Y^c_{n+1} &= Y^c_n + \frac{h}{24}\left(9\,F^p_{n+1} + 19\,F^c_n - 5\,F^c_{n-1} + F^c_{n-2}\right) \end{aligned} \qquad (4)$$

	CYCLE	INPUT	CALCULATION	STORAGE	OUTPUT
$n=0,1,2$	IN	—	—	—	$y^c_{i,0}$
	A_n	Y^c_n, t_n	$f^c_{i,n};\ y^c_{i,n+1}$	$y^c_{i,3}, f^c_{i,n}$	$y^c_{i,n+1}$
	B_m	Y^c_m, t_m	$f^c_{i,m};\ y^p_{i,m+1}$	$f^c_{i,m}$	$y^p_{i,m+1}$
$m=3,\ldots$		Y^p_{m+1}, t_{m+1}	$f^p_{i,m+1};\ y^c_{i,m+1}$	$y^c_{i,m+1}$	$y^c_{i,m+1}$

Fig. 5. DPP81-implementation of the 4-th order PC (Adams-Moulton pair) integration algorithm for the case, that each differential equation is solved in a separate PE: framework of the parallel program being executed in the i-th PE (Capital letters Y refer to vectors; all quantities in the processing cycles IN and An are provided with the superscript c).

in order to solve the system of ordinary differential equations $dY/dt = F(Y,t)$, $Y \in R_n$. In the figure the first column refers to the processing cycles concerned, and the other columns indicate the quantities respectively to be samples via the 16-variables input buffer, to be calculated, to be stored in the data memory for later use in other processing cycles and to be output via the single-variable output buffer. One processing cycle is used to proceed one step in the integration process. Any processing cycle B_m consists of two parts (in the figure separated by a dotted line); between both parts information exchange of the various PE's is affected. The parallel program, executed in the processing cycles B_m, differs from the one in the processing cycles A_n, where e.g. Euler-integration is performed to provide the starting values of algorithm (4). For these non-selfstarting integration methods the DPP81 property of "multiple reconfigurable parallel structuring" can be fruitfully used, which implies that up to and including four different parallel subprograms can be loaded in advance in each PE and software provisions are available for easy switching during run-time from one subprogram to another one to be executed.

The DPP81 has been extensively used among other simulation applications (mainly continuous systems simulations, but also discrete event simulation (Kerckhoffs, 1983; Kerckhoffs et al., 1983; Brok and Kerckhoffs, 1984; Kerckhoffs and Brok, 1985). Meanwhile, research efforts are going on with respect to next DPP-versions (Brok et al., 1983, Andriessen et al., 1984). For example, a DPP84/1/n will consist of one PM (processing module) with a number of n (maximally 16) fully interconnected PE's: the PE's will be equipped with IPP's (inner product processors), which are dedicated processors to perform mappings such as (extended) inner products, transversal and recursive filtering, polynomial mappings, convolutions, auto- and cross-correlations, FFT. Much attention is also paid to parallel programming tools; the research focuses around two main subjects:

1) the construction of a compiler and scheduler for a high level language, where the compiler is meant to transform a

program into a task graph, that indicates the dependencies among the tasks and hence the potential parallelisms, and
2) the development of a hierarchical programming system (Brok et al., 1983). We remark that, unlike the DPP81, the next versions of the DPP will be of the distributed MIMD (DMIMD) type, where the data processing power is distributed over a number of levels.

In a later stage also artificial intelligence techniques are planned to be incorporated into the Delft Parallel Processor project.

4. THE AD10 SYSTEM AND SIMULATION

4.1. Hardware Features

The AD10 is a peripheral multiprocessor system, designed for the simulation of continuous systems. Two factors contribute to the high computational speed, needed in the simulation area:

a) the use of high speed logic circuitry (ECL technology)
b) the incorporation of several architectural features, based on a thorough understanding of the application area (Gilbert and Howe, 1978).

This resulted in a unique design of five different processors, interconnected to one another and to memory by a fast synchronous broadcast bus (50 ns per transaction). The processors are based on pipelining techniques and each one runs an independent program in its own fast memory. A host computer is needed for monitoring and software development, but simulation tasks are performed without its aid.

Some simulation oriented concepts

a) The use of functional processors. A simulation tasks includes structured and unstructured operations, e.g. the numerical integration routines and the non-repetitive computation of right hand side terms of state variable equations. Therefore, dedicated processors are used for efficient handling of the different operations.

- the ARP (Arithmetic Processor) executes fixed point arithmetical instructions of 20 million additions and 10 million instructions per second. It is used for function interpolation and right hand side terms calculations. The basic ARP instruction is of the general form:
  $R = \pm (A \pm B) * C \pm D$
  and is e.g. very suited to compute the linear interpolation equation

  $$f(x) = f_{i+1} - f_i) * D_i(x) + f_i$$

  $$\text{with } D_i(x) = (x - x_i)/(x_{i+1} - x_i)$$

  The ARP is also a good example of the pipeline technique. The instruction is divided into two segments $P = \pm (A \pm B) * C$ and $R = P \pm D$, producing a result in 175 ns. With a filled pipeline a result is obtained each instruction cycle (100 ns) (AD10 Hardware Reference Manual; Gilbert and Howe, 1977,).

- The MAP (Memory Address Processor) and DEP (Decision Processor) efficiently implement the function look up process by special memory addressing techniques and by executing a binary search algorithm. The processors are closely connected by a common index register file.
- The NIP (Numerical Integration Processor) performs the integration of the state variables. It is in fact a sophisticated 48-bit adder with subroutine capability for the implementation of a variety of integration algorithms (Adams-Bashford, Adams-Moulton and Runge-Kutta routines). The 48 bit word length guarantees adequate precision for the simulation of stiff systems (Gilbert and Howe, 1978).
- Finally the COP (Control Processor) has a supervisory function and controls the input/output channels as well.

b) A flexible high speed input/output capability. The AD10 provides for real-time simulation with hardware-in-the loop by an extensive input-output system. This includes analog to digital converters, digital to analog converters, discrete input and output lines and (optionally) a 16 bit parallel digital to digital interface.

c) The AD10 uses 16 bit word length and fixed point format. A 16 bit word length was found to be adequate for the majority of simulation tasks involving empirical data. This also allows a large data memory (up to one million words), needed for multivariant function value storage. Except for the NIP (48 bit), the computer is hence configured around a 16-bit word. Since the prime concern in time critical simulation is the computational speed, most arithmetic operations are performed in fixed point format, as a trade off between system performance and price on the one hand and programmer convenience on the other. To compensate for the need for floating point calculations and accuracy, ADI has recently developed the AD100 system. This is a floating point processor, designed as an extension of the AD10 and with common design concepts. The processor itself is e.g. composed of several functional subprocessors (FX Introduction, ADI Staff, 1984).

d) Parallel processing. The way the different processors work together to perform a simulation task, can be outlined as follows. Each frametime all state variables are updated in the NIP, using a particular (explicit) integration algorithm. At the same time the ARP is calculating right-hand side terms of the differential equations and function table look up is done by MAP and DEP processors. Also, values can be inputted from or outputted to the external world through COP commands. Note that the processors work independently and that timing of data transfers between processors must be taken into account by the programmer.

4.2. Software Features

4.2.1. The MPS10 system

It is obvious from the above description that a software system is needed to make this complex hardware usable. A high level simulation system has been developed by ADI, called MPS10 (Modular Programming System 10). It is a three-file oriented model description system. One file contains the model structure, the second specifies numerical data for a particular simulation run and the third one is a Fortran subroutine that does prerun control and initialization. On top of this a MPS10 preprocessor ("MPSGEN") was built allowing a one-file model description in a more recognizable mathematical form.

The MPS10 system firstly consists of modular programs in AD10 machine code. Each modular program implements a basic computational function like integration, division, function table look up, etc. ... on a number of variables. An MPS10 program describes a model in terms of these basic functions. Each appears to transform a number of inputs into outputs in a quasi parallel way, much like the computing units of an analog computer.

The second part of MPS10 is made up of host computer software. It includes a cross compiler and a control program for interaction with the AD10. Since the AD10 is not a general purpose computer, the host computer is needed for things like software development, interaction with the simulation and program storage. Simulation results are displayed directly on a storage oscilloscope. This approach makes all of the host resources available to the simulation expert and guarantees a good interaction with the simulation. It however involves download times and host-AD10 communication overhead for large problems. Note that a fast host AD10 channel is (optionally) available. It is implemented by the DDC (Digital Device Controller). This device connects the AD10 bus with the DMA channel of the host computer.

4.2.2. Special programming techniques

MPS10 generates fairly efficient code using most of the AD10 features. Speeding up is possible by using a software tool, called "ARP algebra". It is embedded in MPS10 and permits the writing of ARP code in a simplified, assembly-like manner. In this way the special features of the pipelined ARP processor can be fully exploited. It is also useful to implement functions, not available in the standard modular program library.

When the speed really has to be pushed, faster code can also be obtained by making use of the macrofile library In combination with the cross-assembler; thus writing programs in assembly language. This however asks for a very specialized knowledge from the programmer. Arpalgebra will mostly be a more convenient solution.

4.3. An Example: Simulation of a Transmission Line

The AD10's capability of solving large systems of ordinary differential equations is illustrated by a transmission line simulation. It shows the programming of a dynamic system in MPS10 and the performance of the AD10. Also, some "arpalgebra" features are discussed.

4.3.1. The model

The transmission line is modelled by a classic linear lumped representation. At each node, a series inductor (1), a series resistance (r), a parallel capacitor (c) and a parallel admittance (g) represent the distributed parameters of the actual line. The implementation uses 128 of these nodes, resulting in a dynamic system of 256 first order differential equations. Each node is described by:

$$\frac{dI_i}{dt} = c_1 V_i - c_1 V_{i+1} - c_2 I_i$$
$$\frac{dV_{i+1}}{dt} = c_3 I_i - c_3 I_{i+1} - c_4 V_{i+1} \qquad (5)$$

with $c_1 = \frac{1}{l\Delta x}$, $c_2 = \frac{r}{l}$, $c_3 = \frac{1}{c\Delta x}$ and $c_4 = \frac{g}{x}$

r,g,c,l : parameter values per unit of length

Δx : length of a node element

The parameters are considered to have a constant value along the line. This is not a necessity; the line can have variable characteristics along its length without any effect on the programming of the model or on the computational speed.

The model is completed with a (software) voltage generator as excitation of the line and an R-L-C load as termination.

4.3.2. The program

The dynamic system can be programmed in a straightforward way by using the MPS10. Equations (5) are, broadly outlined, programmed in MPS10 as follows:
(for i = 1)
- in the structure description file:

```
INTL    INTDA1    I1    IIC1   3  /  V1   V2   I1  /
INTL    INTDA1    V2    VIC2   3  /  I1   I2   V2  /
```

- in the "specify data" file:

```
PARAMETERS
  C    0.003          L    0.003
  R    5E-04          G    0.
  LEN  3000.
END
```

- in the Fortran subroutine RELATE:

```
DX = LEN/128
C1 = 1./(C*DX)
C2 = R/L
C3 = 1./(C*DX)
C4 = G/C
CALL COEFF('I1',C1,-C1,-C2)
CALL COEFF('V2',C3,-C3,-C4)
```

The "INTL" and "CALL COEFF" sequences must be repeated 128 times to describe the whole model.

Even more simple is the "MPSGEN" preprocessor approach: the price of this approach is a rather long processing time.

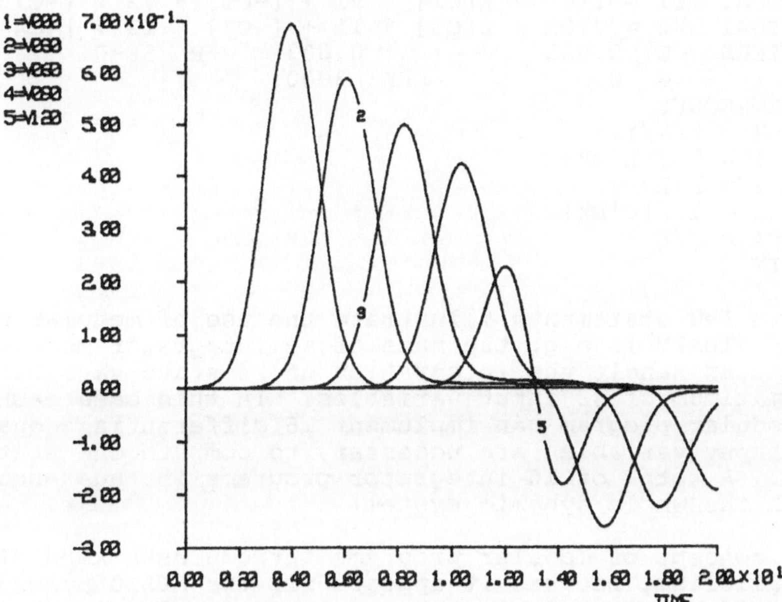

Fig. 6. A gaussian pulse on a short-circuited transmission line: each curve represents a different location on the line.

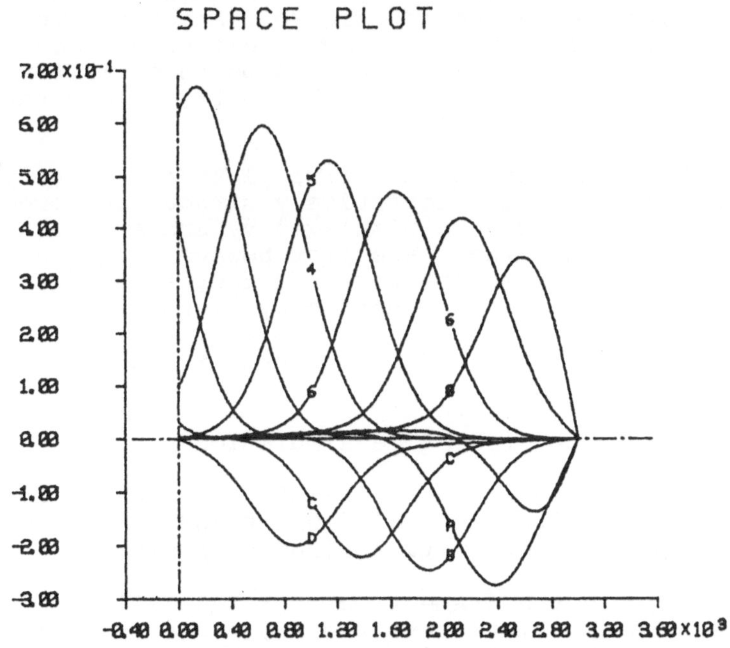

Fig. 7. A gaussian pulse on a short-circuited transmission line: each curve represents a different moment in time.

```
- the MSPGEN program (broadly outlined)/
INT   INTDA1  I1 = IIC1 + $[C1] * V1 + [-C1] * V2 + [-C2] * I1
INT   INTDA1  V2 = VIC2 + $[C3] * I1 + [-C3] * I2 + [-C4] * V2
  PARAMETERS    G  0.003       L   0.003     R  5E-04      %
                G  0.          LEN  3000.
  BEGIN FTNCOEFF
        DX = LEN/128
        C1 = 1./(L*DX)
        C2 = R/L
        C3 = 1./(C*DX)
        C4 = G/C
  ENDBEGIN
```

These few statements illustrate the use of modular programs. "INTDA1" is e.g. the name of an integrator modular program. It can handle the integration of 24 state variables, using a maximum of 48 input variables. In this case each integrator modular program can implement 16 differential equations, since 3 input variables are necessary to compute one state variable. A total of 16 integrator programs is thus enough to cope with the whole dynamic system.

The concept of modular programs introduces a kind of software parallelism, because it appears for the MPS10 programmer as if the 16 state variables are computed simultaneously. As mentioned before, the state variables are not computed physically in separate processing units but the necessary computations are executed in an overlapped way by the different processors (especially ARP and NIP).

Some results are represented in Figures 6 and 7. The voltage generator produces a gaussian pulse and the line is shortcircuited at the end. The pulse is damped along the line by a small resistance value r. Figure 6 shows a time plot of the phenomenon: voltage versus time at different locations along the line. Figure 7 is a space plot that clearly shows the wave travelling. Pictured is the voltage along the entire line at different moments in time.

It should be mentioned that time plots are displayed on the storage oscilloscope interactively without any programming effort. Any variable can be displayed versus another one during a simulation run. The space plot however requires more programming effort and is discussed further in a next paragraph. The generation of the hard copy plots is part of an intelligent software built around the AD10 at the University of Ghent in order to provide for a more user-friendly experimental system.

4.3.3. Performance and real time simulation

The AD10 is designed as a real-time computer. For interfacing convenience with external devices and for maximum speed, the integrations are done with a fixed step size.

In our example the frame time is 1.523 milliseconds. This is the time needed to integrate the 256 state variables. Switching between Adams-Bashford, Adams-Moulton and Runge-Kutta routines while keeping the same frame time can easily be done. This also facilitates synchronizing with real-time events.

Choosing the frame time equal to the step size makes the simulation run in real-time. Since the frame time can mostly be kept small - typical problems range from 50 microseconds to a few milliseconds - phenomena can be simulated in real time with frequencies up to a few hundred Hertz.

Such a real-time solution can be illustrated by the previous model. This can e.g. also represent a travelling pressure wave in a steam-pipe. If the parameters are chosen to simulate a 300 m/s wave travelling in a pipe 3000 m in length, a stable solution is obtained for the maximum step size of:

$5.\ 10^{-4}$	seconds for the	AB1 method	(Adams-Bashford 1st order)
$1.\ 10^{-2}$		AB2	
$2.5.10^{-2}$		AB3	
$2.\ 10^{-2}$		AM2	(Adams-Moulton 2nd order)
$5.\ 10^{-3}$		RK2	(Runge-Kutta 2nd order)

It is seen clearly that for the higher order methods the integration step size can easily be chosen equal to or greater than the frame time ($1.523.10^{-3}$), resulting in a real-time or faster-than-real-time simulation. Assuming 500 integration steps per time constant, pressure changes of 1.3 Hz can be simulated in real-time. Remember that the problem uses 256 state variables.

As far as the accuracy is concerned, the AD10 solution shows a good resemblance to a 64 bit solution with the same number of nodes (Dobbelaere et al., 1985). It is difficult however to increase the accuracy by using a finer mesh, because the number of state variables is limited (theoretically) to 975 (due to the number of available NIP registers).

4.3.4. Using Arpalgebra

The frame time given above applies to the program in its simplest form. During the simulation only a time plot of one state variable can be displayed. Mostly, a complete view of the voltage variation along the entire line is more interesting (Figure 7). Therefore, all of the (128) voltage variables must be displayed each frame time. This requires a software handling. It can be done in standard MPS10 requiring approximately 0.4 millisecond per frame time and a lot of (128) waste variables.

A better approach is the use of Arpalgebra. By inserting Arpalgebra modules in the MPS10 program, it is possible to generate the space plot in 0.12 milliseconds without superfluous variables.

Arpalgebra programs are fast because pipelining and look ahead features of the ARP are fully exploited. Though the code is only sequential (no jump instructions), the experienced programmer can implement logic operations and even "if ... then ... else" structures by making use of round-off features, integer and scaled fraction arithmetic modes of the ARP. The display handling module in our example uses these concepts.

Other applications range from simple switch function modules, over double precision operations (31 bit instead of 16)

(Howe, 1983a), to simulations of six degrees of freedom flight (Howe, 1983b). Mostly, a significant speed-up can be achieved.

As a conclusion, this example has shown the ability of the AD10 as a real time simulation for large dynamic systems. In addition, its function generation capability makes the machine extremely useful in an aerospace engineering and nuclear plant simulation environment. Also, as is now being done at the University of Ghent, it is possible to combine a lot of different models in one program without loss of response time, in order to study (biochemical) systems with poorly known mathematical structures. The simulationist sees and compares the results of several experiments on different models simultaneously.

5. CONCLUSION

In this paper we discussed the application of some multiprocessor systems in continuous systems simulation. In particular, the attention was focused on the EAI SIMSTAR, the Delft Parallel Processor DPP81 and its successors, and the Applied Dynamics AD10 system. A conclusion might be that in some large-scale and complex simulations the usef of non-conventional computers shall be a necessity in order to meet the strong speed requirements. However, the employment of parallel and pipeline computers implies the use of special methods and algorithms to exploit the offered possibilities of parallelism as much as possible. Consequently, the development of appropriate parallel and pipeline algorithms are crucial. A few remarks on them have been made throughout the paper.

REFERENCES

AD10 Hardware Reference Manual, Applied Dynamics International, Ann Arbor, Michigan, U.S.A.

FX Introduction, ADI Staff, Applied Dynamics International, Ann Arbor, Michigan, U.S.A., in: "Proceedings of Adius 84, June 84".

Andriessen, J.H.M., Brok, S.W., Dekker, L., Ruighaver, A.B., and Sips, H.J., 1984, The Delft Parallel Processor/ A distributed MIMD Processor, in: "Proceedings of the 1984 Summer Computer Simulation Conference", W.D. Wade, ed., Society for Computer Simulation, La Jolla, U.S.A., p. 1241-1246.

Barnes, J.H. et al., 1968, The Illiac IV Computer, IEEE Trans. Comp., C17:746-757.

Bekey, G.A., and Karplus, W.J., 1968, "Hybrid Computation", John Wiley & Sons Inc., New York.

Brinton, B., 1982, Systems grows from 10 to 160 MIPS, Electronics (New Products), February 1982.

Brok, S.W., Kooiman, A., de Swaan Arons, H., and Zegwaard, J.F., 1978, in: "Delft Progress Report, Nr. 3", Delft University Press, p. 275-292.

Brok, S.W., Kooiman, A., Llurba, L., Sips, H.J., and de Swaan Arons, H., 1979, The Delft time-shared hybrid minisystem, in: "Simulation of Systems '79. Proceedings of the 9th IMACS World Congress, Sorrento, Italy, 1979", L. Dekker, G. Savastano and G.C. Vansteenkiste, eds., North-Holland Publishing Co., Amsterdam, p. 219-228.

Brok, S.W., Dekker, L., Kerckhoffs, E.J.H., Ruighaver, A.B., and Sips, H.J., 1983, Architecture and programmature of the MIMD-structured Delft Parallel Processor, in: "Proceedings of the First European Simulation Congress ESC 83", W. Ameling, ed., Springer Verlag, Berlin, p. 125-139.

Brok, S.W. and Kerckhoffs, E.J.H., 1984, The Delft Parallel Processor in a simulation environment, in: "Proceedings of the 1984 UKSC Conference on Computer Simulation", D.J. Murray-Smith, ed., Butterworths, London, p. 133-150.

Dekker, L., and Kerckhoffs, E.J.H., 1974, Automatized programming of non-linear differential systems in a hybrid computer/Hybrid simulation of a world model, IMACS (formerly AICA) Journal, October 1974, p. 14-23.

Dekker, L., 1976, Algorithms for parallel processing, in: "Microprocessors and Simulation. Proceedings of the 3rd European Simulation Meeting, Capri, Italy, December 1976", L. Sansone, ed., University of Naples, p. 5-39.

Dobbelaere, B., Bracke, W., and Strybol, L., 1985, The solution of partial differential equations in MPS10, in: "Proceedings of ADIU85, Kissimee, Florida, U.S.A., June 1985".

Embley, R.W., 1984, The technology behind SIMSTAR. An all-new simulation multiprocessor, in: "Simulationstechnik. Proceedings of the ASIM Symposium on Simulation, Vienna, Austria, September 1984", F. Breitenecker and W. Kleinert, eds., Springer Verlag, Berlin, p. 317-332.

Feilmeier, M., 1974, "Hybridrechnen", Birkhauser Verlag, Basel und Stuttgart (in German).

Flynn, M.J., 1972, Some computer organizations and their effectiveness, IEEE Trans. Comp., C-21:948-960.

Franklin, M.A., 1978, Parallel solution of ordinary differential equations, Transactions on Computers, C-27:413-420.

Gilbert, E.O., and Howe, R.M., 1977, An expanded role for function generation in dynamic system simulation, in: "Proceedings of the 1977 Summer Computer Conference, Chicago, U.S.A., July 1977".

Gilbert, E.O., and Howe, R.M., 1978, Design considerations in a multiprocessor. Computer for continuous system simulation, in: "AFIPS Conference Proceedings, Vol. 47".

Hockney, R.W., and Jesshope, C.R., 1981, "Parallel Computers", Adam Hilger Ltd., Bristol.

Hossfeld, F., 1983, Parallele algorithmen, Informatik-Fachberichte, Springer Verlag, Berlin.

Howe, R.M., 1983a, Mathematical modelling and programming techniques for the ADI-system 10, presented at ADIUS 83, Lakeway, Texas, U.S.A.

Howe, R.M., 1983b, ARPALG GDOF Aircraft Module, Applications Report, ADI, Ann Arbor, Michigan, U.S.A.

Ilid, Z.V., 1984, SIMSTAR/Application areas survey, in: "Simulationstechnik. Proceedings of the ASIM Symposium on Simulation, Vienna, Austria, September 1984", F. Breitenecker and W. Kleinert, eds., Springer Verlag, Berlin, p. 611-613.

Johnson, P.M., 1978, An introduction to vector processing, Computer Design, 17:89-97.

Karplus, W.J., 1977, Peripheral processors for high-speed simulation, Simulation, 29:143-153.

Kerckhoffs, E.J.H., Dekker, L., and van Gelderen, J.A., 1973, A block-iteration method for the hybrid solution of large

systems of ordinary differential equations, in: "Proceedings of the 7th AICA Congress, Prague 1973", p. 125-129.
Kerckhoffs, E.J.H., 1982, The use of parallel overlays in hybrid computer simulation, in: "Proceedings of the International AMSE Conference on Modelling and Simulation, Paris, France, July 1982", G. Mesnard, ed., AMSE, Volume 3, Group 3 (Simulation Methods), p. 65-69.
Kerckhoffs, E.J.H., 1983, The application of an experimental parallel processor for the simulation of systems in biotechnology and medical engineering, in: "Modelling and Data Analysis in Biotechnology and Medical Engineering, G.C. Vansteenkiste and P.C. Young, eds., North-Holland Publ. Co., Amsterdam, p. 221-231.
Kerckhoffs, E.J.H., and Brok, S.W., 1985, "The Delft Processor DPP81: properties and utilization in simulation and related fields systems analysis, Modelling and Simulation (Journal of Mathematical Modelling and Simulation in Systems Analysis), Akademie Verlag, Berlin, 2:175-208.
Kerckhoffs, E.J.H., Potucek, J., and Snorek, M., 1983, Simulation of myocardial fibre action potential generation units on the Delft Parallel Processor, in: "Proceedings of the First European Simulation Congress ESC 83", W. Ameling, ed., Springer Verlag, Berlin, p. 138-148.
Korn, G.A., and Korn, T.M., 1964, "Electronic Analog and Hybrid Computers", McGraw-Hill, New York.
Kuck, D.J., 1968, Illiac IV software and application programming, IEEE Trans. Comp., C-17:758-770.
Landauer, J.P., 1983, SIMSTAR - An attached multiprocessor for dynamic system engineering, in: "Proceedings of the First European Simulation Congress ESC 83", W. Ameling, ed., Springer Verlag, Berlin, p. 155-171.
Louie, T., 1981, Array processor: a selected bibliography, Computer, September 1981, p. 53-57.
Nievergelt, J., 1984, Parallel methods for integrating ordinary differential equations, Communications of the ACM, 7:731-733.
Miranker, W.L., and Liniger, W.M., 1967, Parallel methods for the numerical integration of ordinary differential equations, Mathematics of Computation, 21:303-320.
SIMSTAR Technical Description, report of Electronic Associates Inc., 1983.
Spriet, J.A., and Vansteenkiste, G.C., 1982, "Computer Aided Modelling and Simulation", International Lecture Notes on Computer Science, Academic Press, London.
Worland, P.B., 1976, Parallel methods for the numerical solution of ordinary differential equations", IEEE Trans. on Comp., October 1976, p. 1045-1048.
Yen, K., and Cook, G., 1982, Digital simulation algorithms using parallel processing, IEEE Transactions on Industrial Electronics, IE-29:217-219.
Yura, E., Yoshikawa, R., Nara, Y., Kimura, T., and Aiso, H., 1975, An approach to parallel processing for continuous dynamic system simulation with microprocessors, in: "Proceedings of the 2nd USA-Japan Computer Conference 1975", p. 8/1/1-8/1/6.

APPENDIX: THE DELFT PARALLEL PROCESSOR

The machine's properties can shortly be summarized as follows:

1) Number of processing elements (PE's): 8
2) At any time each PE can perform an arithmetic operation from an extensive set of operations B = {+,-.,,/,...}. Different operations require different lengths of time.
3) At any time, different PE's can perform different operations (from the set B).
4) Full interconnectability between PE's. Each PE can communicate with the host computer.
5) In the so-called synchronous mode of operation, a "processing cycle" is characterized by a common start of all PE's, and is ended as soon as all (!) PE's have finished their program execution. Information exchange between the PE's is always affected at the very end of each processing cycle, but can also be affected a number of times during a processing cycle. In continuous systems simulation one or more processing cycles are used to proceed one step in the integration process.
6) Each PE has a single-valued output and can sample at its input 16 variables.
7) "Reconfigurable parallel structuring" allows loading of four subprograms in each PE and easy switching from one subprogram to another one during run time.

THE NUMERICAL SOLUTION OF ELLIPTIC PARTIAL DIFFERENTIAL EQUATIONS ON A HYPERCUBE MULTIPROCESSOR

S. Vandewalle, J. De Keyser, and R. Piessens

Katholieke Universiteit Leuven
Departement Computerwetenschappen
Celestijnenlaan 200A
B-3030 Leuven

e-mail: stefan@kulcs.{UUCP,BITNET}

I. INTRODUCTION

The numerical solution of partial differential equations is a computationally very demanding problem. Many researchers have therefore devoted their efforts to the implementation and analysis of existing methods on parallel machines as well as to the development of innovative numerical techniques[3,4,13,15]. The number of articles, reports and communications on the subject is impressive, indicating its importance in current science and engineering[16].

In this paper we present a detailed discussion on the implementation of a number of classical iterative methods for solving elliptic partial differential equations on the Intel iPSC®/2, a hypercube multiprocessor. We discuss a library of solvers for the general class of second order, linear, elliptic PDE's with Dirichlet, Neumann, mixed or periodic boundary conditions on a rectangle.

In section 2 we discuss the hypercube topology and its important characteristics. The problem class and its discretization are presented in section 3. In section 4 we explain the basic parallel solution scheme. The implementation of several classical iterative methods, Jacobi, Gauss-Seidel, SOR, Conjugate Gradients and Multigrid, is discussed in section 5. In the final section we present a test problem and timing results obtained on an Intel iPSC/2.

II. HYPERCUBE PARALLEL PROCESSORS

A. Introduction

Multiprocessors can be divided into two classes. In *shared memory* machines processors communicate through access to a common memory. A fundamental drawback is that they are not easily scalable. Limited system bus capacity and memory access collisions prohibit the use of a very high degree of parallelism. In *distributed memory* machines each processor possesses a local memory and messages are passed over an interconnection network. They do not suffer from the unscalability problem. When the number of processors increases so will the number of communication links and the communication bandwith. It is generally agreed upon that the orders of speed improvement needed in scientific computations can only come from this type of multiprocessor.

An important concept is the *grain-size* of a distributed-memory parallel machine. Machines with a fine grain-size consist of many simple processors working together. The functionality of the communication system is normally limited and can therefore be relatively fast. Coarse grain-size machines have relatively few but powerful nodes, possibly equipped with a floating-point accelerator or a vector board. Communication between the processors is typically expensive and should therefore occur rather infrequently. Many commercially available hypercube multiprocessors are of the latter type. They are general-purpose, distributed-memory, coarse grain-size parallel machines, the processors of which are connected in a hypercube topology.

B. The hypercube topology

The hypercube structure is generally defined in a recursive manner[18,4], see figure 2.1. A hypercube of dimension 0 consists of one node. A hypercube of dimension d consists of 2 hypercubes of dimension d-1 whose corresponding nodes are connected.

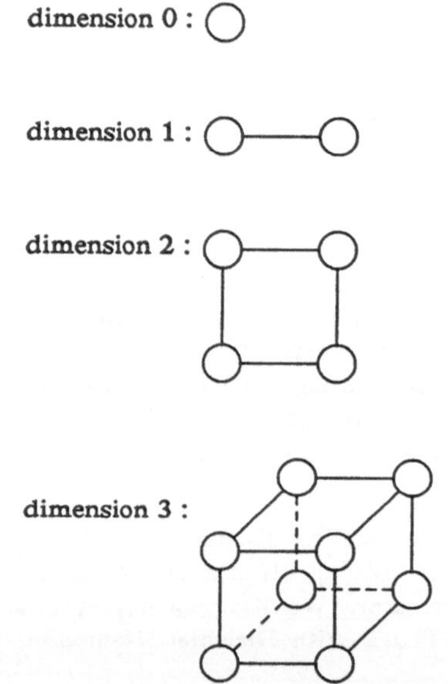

Figure 2.1. Recursive definition of the hypercube topology

Hypercube networks have a number of special characteristics:

- The total number of nodes is 2^d, if d is the cube dimension.
- The diameter of a d-cube is d, i.e. the distance between any two nodes is at most equal to the 2-logarithm of the number of nodes.
- The degree of the network, which is defined as the maximum number of edges emanating from a node, equals d.
- The nodes of a d-dimensional hypercube can be numbered by d-bit binary numbers in such a way that there is a physical connection between two nodes if and only if their binary representation differs by one and only one bit (see figure 2.2).
- The distance between two nodes equals the number of different bits in their bit representation.

- Using properties of binary reflected Gray codes, one can easily prove that various other topologies can be mapped onto a hypercube. See figure 2.3 for a mapping of a ring, a tree and an array on a three dimensional cube.

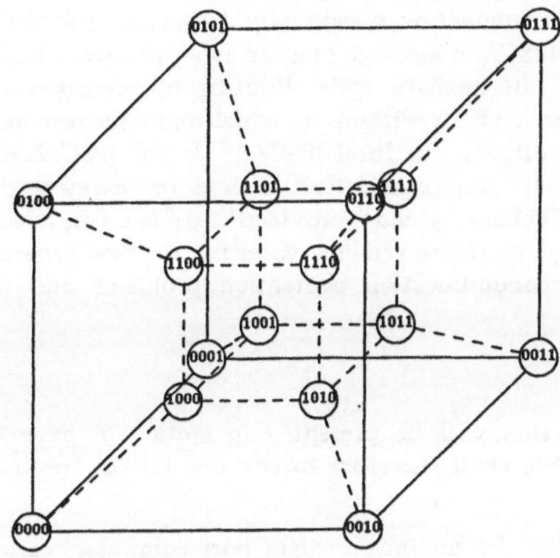

Figure 2.2. A 4-dimensional hypercube with binary numbering

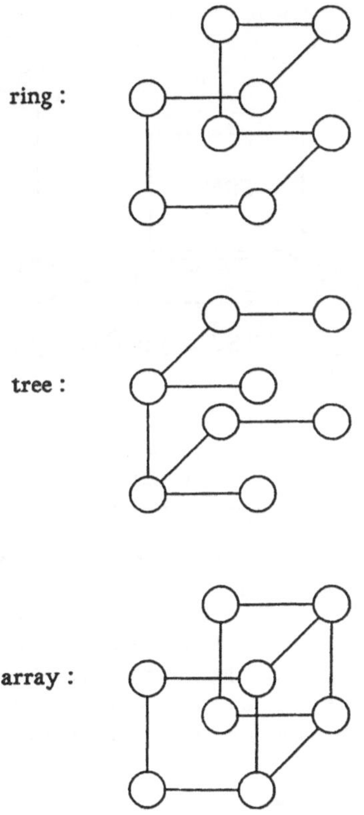

Figure 2.3. Mapping of ring, tree and array on a hypercube of dimension three

The rich connectivity of the hypercube topology has made it to be the interconnection structure for multiprocessors favoured by many hardware manufacturers. The processors are thought to lie on the corners of the multi-dimensional cube, the communication links along the edges. As was mentioned before most communication structures can be mapped onto the hypercube in such a way that only *nearest-neighbour* communication is necessary. This is a favourable situation for the speed of message exchange as well as for the communication link contention. The former argument was especially important for the previous generation of hypercubes. On the Intel iPSC/1 a message to a far away processor had to be received, buffered and forwarded in each intermediate node. Routing of messages was handled purely in software. The development of algorithms in which only nearest-neighbour communication was needed, had full priority on the Intel iPSC/1. On the iPSC/2 each node has a separate message handling processor. Communication with a far away node is almost as fast as communication with a neighbouring node, provided there is a free communication path to that node. The latter argument therefore remains valid for the new generation of hypercubes. By using nearest-neighbour communication contention problems and network saturation are avoided.

C. The Intel iPSC/2

The timing results that will be presented in section 6, have been obtained on a 16 processor Intel iPSC/2. We shall therefore briefly discuss the specific characteristics of that machine.

The cube is controlled by an intermediate host computer, called the *System Resource Manager* or *SRM* (see figure 2.4). Its processor is an Intel 80386 running at 16Mhz with a 80387 mathematical coprocessor.

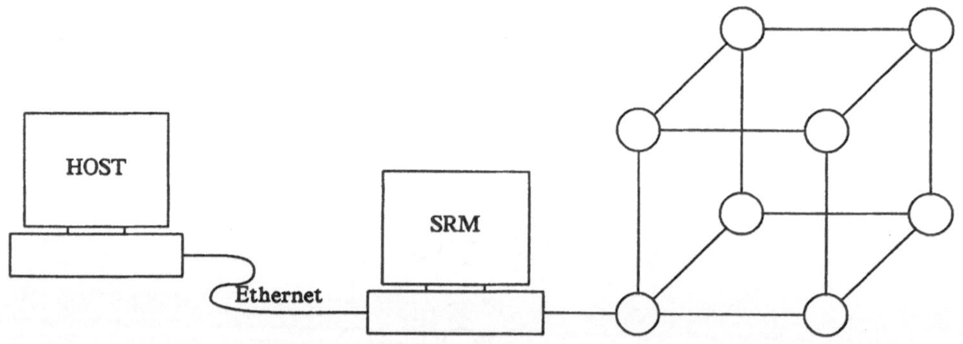

Figure 2.4. Intel iPSC/2 system configuration

It operates the software development tools for the cube programs: compilers, verifiers, loader, diagnostics and a concurrent debugger. It allows *cube sharing*. Several users can simultaneously operate on subcubes. The SRM currently runs Unix^{TM} System V. It can be connected with a network of workstations and minis, which are called the *host* computers.

Each processor board of the cube has a 80386/87 processor, 4Mbyte of memory (maximum configuration has 16 Mbyte per node) and a separate communication processor called the *Direct Connect Module* or *DCM*. The DCM supervises 8 full duplex serial channels with performance of 2.8 Mbytes per second and per channel. Message routing and the set-up of the communication path is done by the hardware. Multi-hop messages (messages to non-neighbour processors) do not interfere with the computation going on in the intermediate nodes. The nodes run a small operating system, the Node eXecutive/2 or NX/2.

Unix is a trademark of AT&T Bell Laboratories.

The programmer normally writes one or more node programs which are distributed to an allocated subcube. On each node several processes can be running. A host program takes care of the I/O operations such as the input from the keyboard and the output to the screen or graphics device. Some of the I/O operations can also be performed from the nodes. This is especially helpful while debugging the code. The host program can run on the SRM or on one of the hosts. In the latter case the operation of the SRM is totally transparent to the user. Commands are provided to the user to operate the cube from his usual environment (e.g. Unix 4.3Bsd on a workstation).

D. The communication system

Communication on the iPSC/2 hypercube is *asynchronous*. There is no rendez-vous between processors at the moment of message passing. When the destination processor is not yet expecting a message at the time of its arrival, it is buffered by the node operating system.

The messages can be *typed*. They can be given a label specifying a type, which gives an indication of the nature of contents of the message. The receiver can select a message by specifying the type of message he wants to receive. This selection can be suppressed, indicating that the receiver will accept any message of whatever type.

The communication primitives exist in a *blocking* and *non-blocking* version. In the first case, the sending or receiving process is halted until the message has been sent or received. In the case of a send operation the buffer that contained the message is then free for further use. In the case of a receive operation the message is then available in the buffer specified by the programmer. In the non-blocking operation the program merely informs the operating system that a message should be sent or received. The processor is allowed to proceed with computations while the communication processor handles the message request. The communication buffer should not be accessed or re-used until the process is informed by the operating system about the end of the communication operation.

In table 2.1 we present some of the basic communication primitives provided by the communication system of iPSC/2[11]. More involved primitives such as broadcasts and global operations are available in a subroutine library.

Table 2.1: basic communication primitives

type	syntax	description
blocking	csend(type,buf,len,node,pid)	send a message and wait for completion
	crecv(type,buf,len)	receive a message and wait for completion
non-blocking	int isend(type,buf,len,node,pid)	send a message
	int irecv(type,buf,len)	receive a message
	int msgdone(id)	determine whether communication has completed
	int msgwait(id)	wait until completion of communication operation

As an example we shall explain the parameters of the non-blocking send operation *isend*. The destination process is identified by its process number *pid* and the nodenumber *node* of the processor on which it runs. The message to be sent has type *type*, is contained in the buffer *buf* and has the length *len*. On return of the routine an identification is given for future reference to the message. The calling process can now continue its operation while the message is being sent. The buffer should not be re-used until the process is informed by means of a *msgdone* or *msgwait* instruction that the send operation has finished.

E. Some timing results

1. Computation speed

In fig. 2.5 we present timing results obtained for some double precision floating point operations.

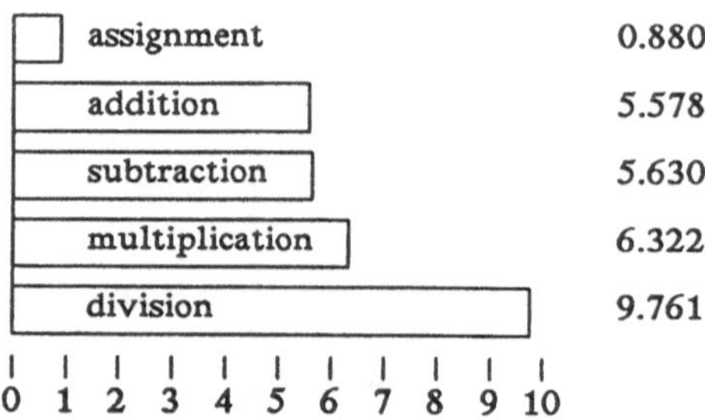

Figure 2.5. Timing results for double precision floating point operations (*μsec*)

For more complicated sequences of operations the speed somewhat increases due to a certain amount of pipelining in the 80387 mathematical coprocessor.

2. Communication speed

In figure 2.6 we present timing results for a nearest neighbour send-and-reply. This communication operation is basically performed in the following way. A message is sent from node 0 to node 1. It is received by node 1 and reflected back to node 0. After receipt of the message by node 0, the operation is completed. The total time for this send-and-reply is divided by two to get the time needed to transmit a message between two neighbouring nodes. A more detailed discussion on how the timings were obtained as well as a number of other timing results can be found elsewhere[1]. As can be seen from figure 2.6 different communication strategies are used for small messages (less than 100 bytes) and long messages (more than 100 bytes).

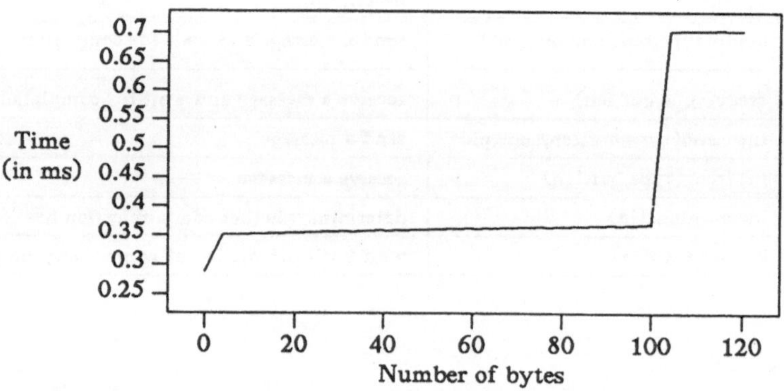

Figure 2.6. Nearest neighbour Send-and-Reply, message lengths from 0 to 120 bytes, multiples of 4 bytes, time per csend-crecv

Communication can be modeled using the formula:

$$t(n) = t_{startup} + n\, t_{send}, \tag{2.1}$$

with n the number of bytes transferred. Different values for $t_{startup}$ and t_{send} are found for short and long messages, see table 2.2.

Table 2.2: communication parameters for Intel iPSC/2		
	$t_{startup}$	t_{send}
short messages	350 μsec	0.2 μsec
long messages	660 μsec	0.36 μsec

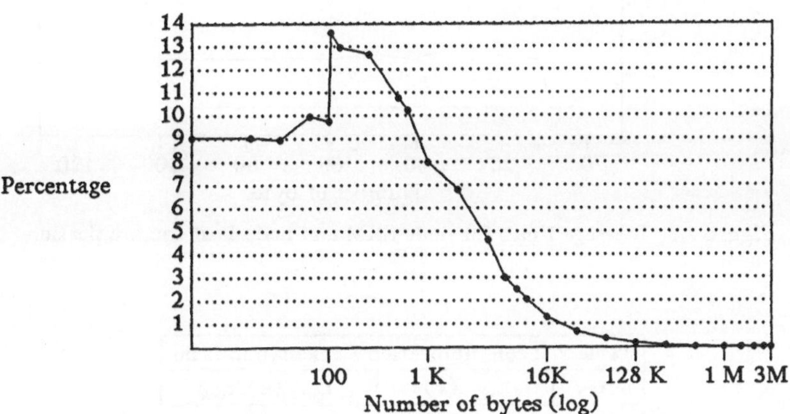

Figure 2.7. Multi-hop Send-and-Reply, message lengths from 0 to 3 M bytes, percentage overhead sending over 4 hops, compared to sending to neighbour

On the iPSC/2 message traffic between non-neighbouring nodes is handled by the Direct Connect communication processors of the sender, receiver and intermediate nodes. The node processors of the intermediate nodes are not affected. In figure 2.7 the percentage overhead of a 4-hop send-and-reply operation versus a nearest-neighbour communication is presented for different values of the message length. The maximum relative overhead is about 14% for small messages and quickly decreases for longer messages.

The major bottleneck of the Intel iPSC/1 hypercube was the very slow communication speed between a host and the cube. It was not uncommon for the output of the results to take much longer than the actual computations. The situation has been seriously improved for the iPSC/2 but there is still a substantial overhead associated with the communication from a node to the System Resource Manager or to a host, see figure 2.8.

F. Some definitions and further considerations

Several fundamental performance parameters can been defined[4].

— t_{calc} : is the time required for one (4 byte) floating point operation

— t_{comm} : is the time required for a 4 byte-word communication between adjacent processors.

The second parameter is very dependent on the message length, due to the very long $t_{startup}$. The definition is therefore adapted and becomes $t_{comm}(k)$: the time per 4 byte word to send a message of k bytes. The ratio $t_{comm}(k)/t_{calc}$ occurs in many theoretical performance calculations.

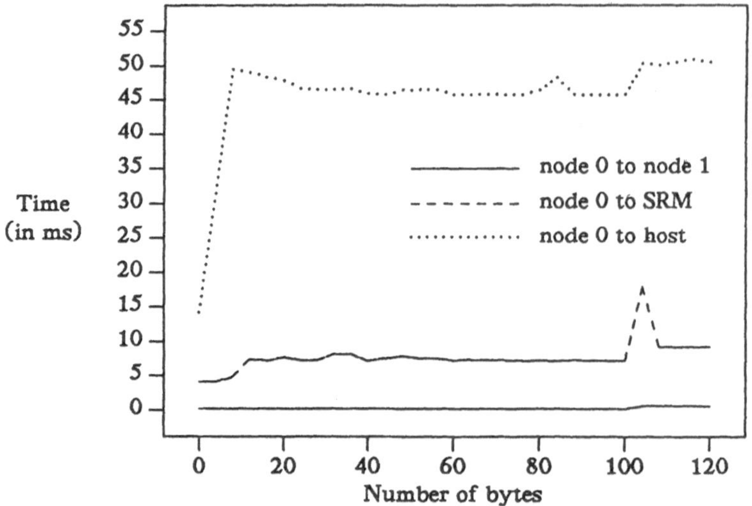

Figure 2.8. Message times for node-SRM and node-host communication

Table 2.3: communication / calculation ratio		
bytes	$t_{comm}(k)$	$t_{comm}(k)/t_{calc}$
4	350	70
8	175	35
100	15	3
104	27	5.4
1024	4	0.8

In table 2.3 we have tabulated the values of this characteristic parameter for different values of k. For very short messages the communication time is dominated by $t_{startup}$. The ratio is very high, showing an important imbalance between communication and calculation speed. For long messages the system is well-balanced. It is therefore very important to structure the algorithms in such a way that only long messages are sent.

The *efficiency* of a concurrent program is defined as:

$$\epsilon = \frac{t_{seq}}{t_p \cdot p} = \frac{S_p}{p} \tag{2.2}$$

with p being the number of processors. The time t_p is the execution time of the parallel program on p processors. The value t_{seq} is the execution time of the best sequential algorithm on one processor. S_p is called the *speedup* on p processors.

Definition 2.2 is rather impractical. It is not always clear what the best sequential algorithm is. Moreover a good implementation of that algorithm is not always available. The definition for ϵ is therefore usually replaced by

$$\epsilon = \frac{t_{parallel\ on\ 1\ processor}}{p\ t_{parallel\ on\ p\ processors}} \tag{2.3}$$

This is a measure for the overhead incurred by the parallel organization of the algorithm. It is due to communication and load-imbalance.

III. PROBLEM CLASS AND DISCRETIZATION

A. Problem class

The routines are written to handle the fairly general class of *linear, second-order, variable coefficient, elliptic* partial differential equations:

$$C_{xx}(x,y)\frac{\partial^2 u}{\partial x^2}+C_{xy}(x,y)\frac{\partial^2 u}{\partial x \partial y}+C_{yy}(x,y)\frac{\partial^2 u}{\partial y^2}+C_x(x,y)\frac{\partial u}{\partial x}+C_y(x,y)\frac{\partial u}{\partial y}+C(x,y)u = f(x,y) \tag{3.1}$$

on a rectangular domain $\Omega = [a,b]x[c,d]$, with boundary conditions of type

— Dirichlet: $u(x,y)=g(x,y) \quad on \ \partial\Omega$

— Mixed: $\frac{\partial u(x,y)}{\partial n} + r(x,y)u(x,y) = s(x,y)$

— Periodic: $\begin{cases} u(a,y) = u(b,y) \\ \frac{\partial u}{\partial x}(a,y) = \frac{\partial u}{\partial x}(b,y) \end{cases}$ or $\begin{cases} u(x,c) = u(x,d) \\ \frac{\partial u}{\partial y}(x,c) = \frac{\partial u}{\partial y}(x,d) \end{cases}$

The derivative in the expression for the mixed condition is taken in the outward normal direction.

B. The finite difference discretization

1. discretization of the differential equation

Equation 3.1 and the boundary conditions are discretized by using finite differences on an equidistant grid. The grid spacing in the x-direction equals h; the grid spacing in the y-direction equals k.

The elliptic partial differential equation is discretized by using central differences:

$$\left[\frac{\partial^2 u}{\partial x^2}\right]_{ij} = \frac{u_{i+1,j} - 2u_{i,j} + u_{i-1,j}}{h^2} \tag{3.2}$$

$$\left[\frac{\partial^2 u}{\partial x \partial y}\right]_{ij} = \frac{u_{i+1,j+1} - u_{i-1,j+1} - u_{i+1,j-1} + u_{i-1,j-1}}{4hk} \tag{3.3}$$

$$\left[\frac{\partial u}{\partial x}\right]_{ij} = \frac{u_{i+1,j} - u_{i-1,j}}{2h} \tag{3.4}$$

Similar formulae are used for discretizing $\frac{\partial^2 u}{\partial y^2}$ and $\frac{\partial u}{\partial y}$.

For each gridpoint this results in an algebraic equation with the following general nine point stencil:

$$
\left|\begin{array}{ccc}
\dfrac{-C_{xy}(x_i,y_j)}{4kh} & \dfrac{C_y(x_i,y_j)}{2k}+\dfrac{C_{yy}(x_i,y_j)}{k^2} & \dfrac{C_{xy}(x_i,y_j)}{4kh} \\[2ex]
\dfrac{-C_x(x_i,y_j)}{2h}+\dfrac{C_{xx}(x_i,y_j)}{h^2} & C(x_i,y_j)-2(\dfrac{C_{xx}(x_i,y_j)}{h^2}+\dfrac{C_{yy}(x_i,y_j)}{k^2}) & \dfrac{C_x(x_i,y_j)}{2h}+\dfrac{C_{xx}(x_i,y_j)}{h^2} \\[2ex]
\dfrac{C_{xy}(x_i,y_j)}{4kh} & \dfrac{-C_y(x_i,y_j)}{2k}+\dfrac{C_{yy}(x_i,y_j)}{k^2} & \dfrac{-C_{xy}(x_i,y_j)}{4kh}
\end{array}\right| \tag{3.5}
$$

When there are no cross derivatives in equation 3.1, i.e. $C_{xy}(x,y)\equiv 0$, the stencil reduces to a five point star.

2. discretization of the boundary conditions

a. Dirichlet boundary conditions.

There are no unknowns associated with gridpoints on boundaries satisfying a Dirichlet condition, since their value is exactly known. They are eliminated from the system of equations by adapting the stencil coefficients and the right hand sides associated with neighbouring gridpoints.

b. Mixed boundary conditions.

The boundary condition is discretized by using a second order correct central difference formula, which relates an interior point and the boundary point to a non-existing exterior point, see figure 3.1. The value at that exterior point can be eliminated by combining the formulae of the discretized PDE with

$$\frac{u_{exterior} - u_{interior}}{2h} + r(x_i,y_j)\, u_{border} = s(x_i,y_j) \tag{3.6}$$

Figure 3.1. Discretization of a mixed boundary condition

c. Periodic boundary conditions

For second-order problems, periodic boundary conditions imply the equality of the solution and its first derivative on opposite borders. This type of boundary condition is easily dealt with. There are no changes in the stencil, nor in the right hand side. The gridpoints are chosen in the manner presented in figure 3.2.

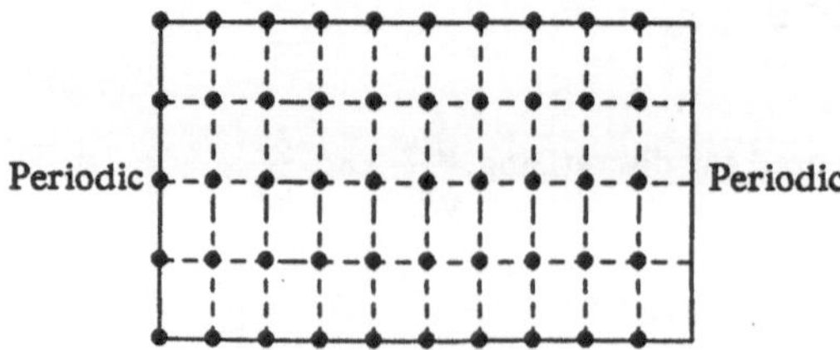

Figure 3.2. Choice of gridpoints in case of periodic boundary conditions

d. Boundary conditions at the corners

When using a nine-point stencil, special attention is needed at the corner points of the domain.

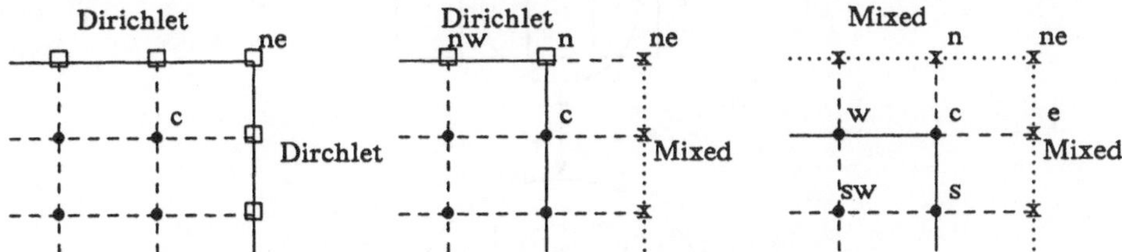

• : interior unknowns, to be determined

□ : known Dirichlet values, to be eliminated

x : exterior unknowns, to be eliminated

Figure 3.3. Gridpoints near domain boundaries

Several combinations of boundary types are possible, see figure 3.3. We shall discuss three cases: a Dirichlet-Dirichlet corner, a Dirichlet-mixed and a mixed-mixed corner. In each case the unknown u_{ne} has to be eliminated from the equation for the corner unknown u_c.

— When two Dirichlet borders meet, the value of u_{ne} is taken to be the mean value of the dirichlet function for the north and east boundary. It is easily eliminated since its value is known.

— In the second case u_{ne} can be determined by expressing the mixed condition at the cornerpoint n. This gives an equation in three unknowns u_{nw}, u_n and u_{ne}. Since the first two variables get their values from the Dirichlet condition, u_{ne} is readily determined.

— In the mixed-mixed case the ficticious gridpoint is eliminated from the stencil of u_c by adding an additional equation, based on the boundary conditions:

$$\frac{\partial u}{\partial n_{sw-ne}} = \nabla u \, \vec{n}_{sw-ne} \quad \text{with } \vec{n}_{sw-ne} \text{ the (normalized) diagonal direction.} \qquad (3.7)$$

IV. THE PARALLEL SOLUTION METHOD

A. Decomposition of the domain

Domain decomposition is applied as proposed by several authors[15,4,10] and presented in figure 4.1. The processors are arranged into a rectangular array structure and mapped onto the computational domain. Each processor is responsible for all computations on the grid points in the interior of the allocated subrectangle and on its south and west boundaries. Since the sizes of the subdomains are equal, an almost balanced load distribution will be guaranteed.

The neighbouring gridpoints are involved in updating the value at a given point. The equation associated with a gridpoint on the border of a processor's subdomain thus involves one or more values at gridpoints allocated to other processors. In order to avoid sending a message each time such a value is referenced, the neighbouring gridlines are stored as well. In the sequel we shall refer to these as the *gridlines in the overlap area.* The overlap area and the local subdomain together constitute the *extended subdomain* for which the processor is responsible. In figure 4.1 the extended region of one processor is shown enlarged. If higher-order differential equations had to be discretized, more duplicate gridlines would have to be stored.

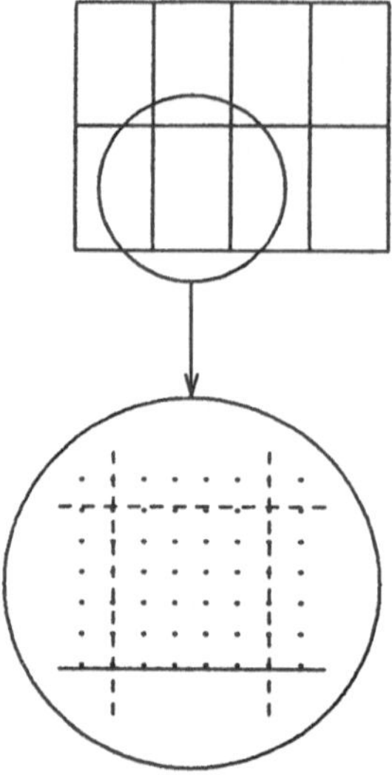

Figure 4.1. Domain decomposition

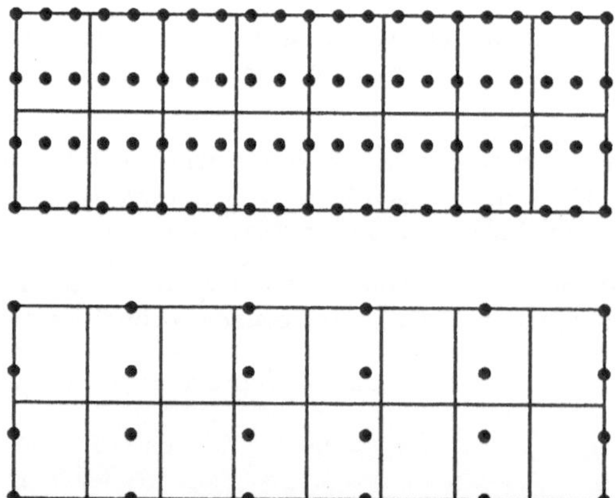

Figure 4.2. Active and inactive processors

We call a grid *consistent* if the gridpoints in the overlap area have their correct values, i.e. the values of the corresponding gridlines in the neighbouring processors. Some operations on grids only affect or use interior unknowns. The grids they operate upon needn't be consistent. The use of inconsistent grids then reduces communication time.

The possibility of *inactive* processors, i.e. processors which have no grid points in their subdomain, is explicitly provided and taken care of in our implementation. No restriction is imposed on the number of equations versus the number of processors. This strategy will be important in the discussion of the multigrid method. In figure 4.2 the decomposition of two grids is shown for a two by eight processor mesh. Each processor is *active* on the first grid. Each processor has at least one gridpoint in its subdomain. Some processors have no gridpoints allocated on the second grid. They will not contribute to computations on this grid.

B. Basic solution scheme

In fig 4.3 we outline a possible parallel program scheme, using library routines as basic building blocks. The dashed arrows indicate communication between host and node processors. Communication between different node processors occurs during the operations put into a box with double-line sides.

The host program is started on a host computer or on the System Resource Manager. It loads the node program in each processor of the subcube allocated to the user. A number of problem parameters, such as dimension of the domain, number of discretization lines in x and y direction, algorithm specific parameters etc..., are entered by the user and sent to the nodes. The host program now enters a loop and starts receiving intermediate results such as error or residual norms.

The node programs are loaded by the host and immediately start execution. They receive some problem and algorithm specific parameters. They calculate the dimensions of their data-structures. After dynamic creation and initialization the main computational loop is entered. Using interprocess communication a sequence of approximations is generated using one of the iterative algorithms. After each iterate some error or residual norm is calculated and sent to the host. After convergence or after a given maximum number of iterations have been done, the results are transmitted to the host.

If the output is desired in graphical form, contourlines will be computed. The processor first communicates to determine the extremal values of the solution. Then the contourlevels are computed. Without any further internode communication each processor calculates the contour lines in its local subdomain and sends them to the host program.

V. THE ITERATIVE ALGORITHMS

A. Introduction

After discretization of the partial differential equation and incorporation of the boundary conditions a system of linear equations results,

$$A\,x = b \qquad \text{with } A = L + D + U \tag{5.1}$$

where A is the sparse coefficient matrix and L, D and U hold the elements below, on and above the diagonal. In the following sections we shall briefly review some of the iterative solution methods[6] and comment on their parallel implementation.

B. The Jacobi method

The Jacobi method consists of applying the iterative scheme,

$$D\,x^{(k+1)} = b - (\,L + U\,)\,x^{(k)} \tag{5.2}$$

repeatedly to an arbitrary starting solution. Convergence is guaranteed whenever the spectral radius $\rho(-D^{-1}(L+U)) < 1$; this requirement is fulfilled for strictly diagonal dominant matrices. The actual convergence rate depends on the problem. It can be shown however that the convergence rate quickly approaches unity when the matrix dimension grows. As the number of gridpoints increases, the convergence rate becomes very low.

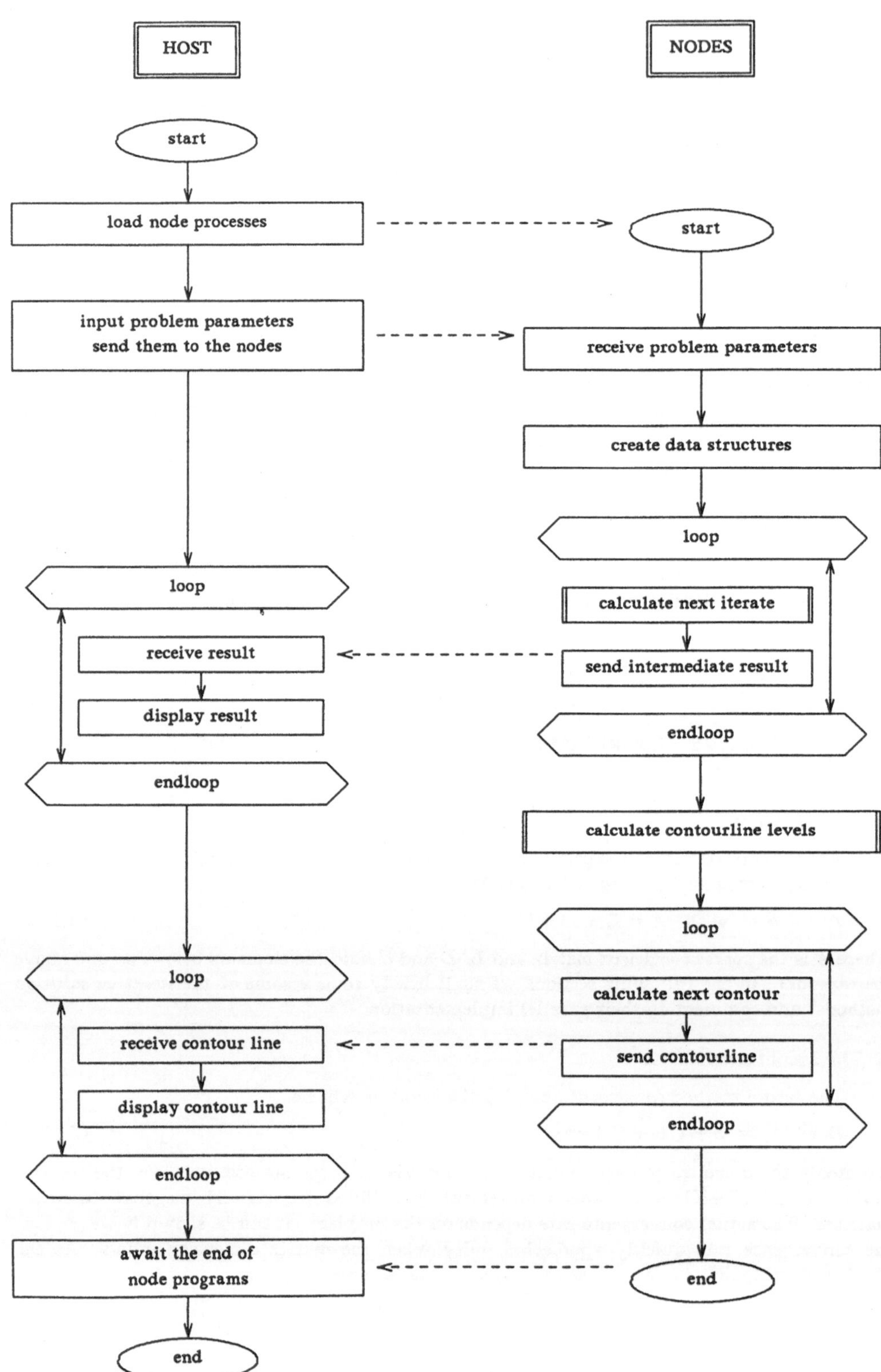

Figure 4.3. Basic program scheme

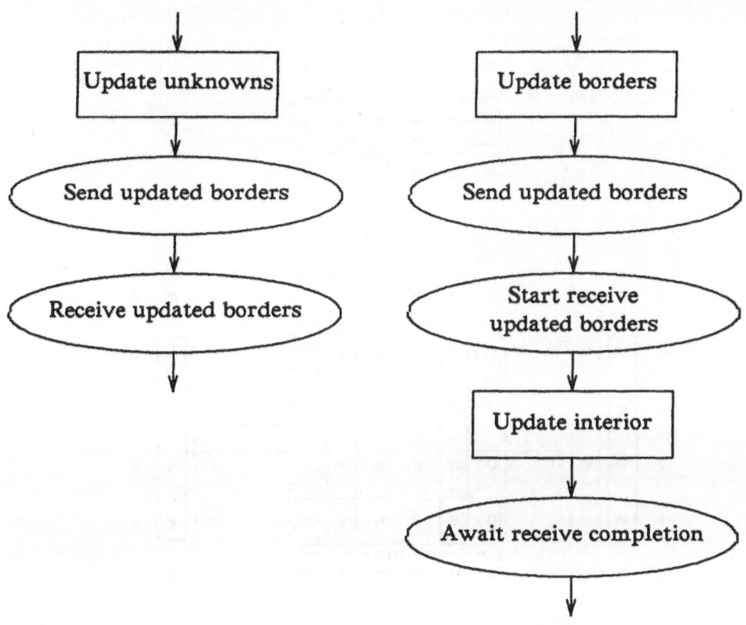

Figure 5.1. a. Simple algorithm b. Improved algorithm

The result of a Jacobi iteration is independent of the order of updating. Since only information of a previous iterate is used, all gridpoints can be updated concurrently. A very simple parallel algorithm is presented in figure 5.1.a. Each processor calculates the new values of the gridpoints in its subdomain. Then the values in the overlap area are updated by exchanging the border gridlines. We call this operation the *flush*-operation. During a flush, information is sent to the four active neighbours and information is received from the same processors. The exchange of information for a five-point-stencil-grid is depicted in figure 5.2. The gridpoints depicted with "•" are gridpoints that belong to the interior of a processor's domain. The "o" gridpoints belong to the overlap region.

This very simple scheme has a significant drawback. The calculations are halted during the communication process. Valuable computation time is wasted while waiting for message traffic to finish.

On the Intel iPSC/2 communication can actually overlap calculation. In figure 5.1.b we present an algorithm that takes this into account. While a message is being sent or received, the program can continue. In case of a five-point stencil this overlapping can be accomplished in the following manner. First the border points of the local subdomain are updated. Immediately afterwards we initiate the send of the four borders to the neighbouring active processors (if the previous send in that direction has finished). All sends proceed in parallel. While the send is in progress, we initiate the receipt of the borders and continue with the calculation of the interior unknowns. If the number of gridlines is sufficiently large, the receive operation will have completed before the calculations terminate.

A somewhat more complex scheme has to be used for the exchange of information when a nine-point-stencil is used. Each processor knows the identity of its neighbours to the north, east, south and west. It does not know the identities of the processors in the diagonal directions. To insure correctness of the corner gridpoints in the overlap region the communication scheme depicted in figure 5.3 is implemented. A horizontal flush is followed by a flush in the vertical direction. Thus some sequentialism is explicitly introduced in the algorithm. Again partial overlap of communication and calculation can be achieved, as shown in figure 5.4.

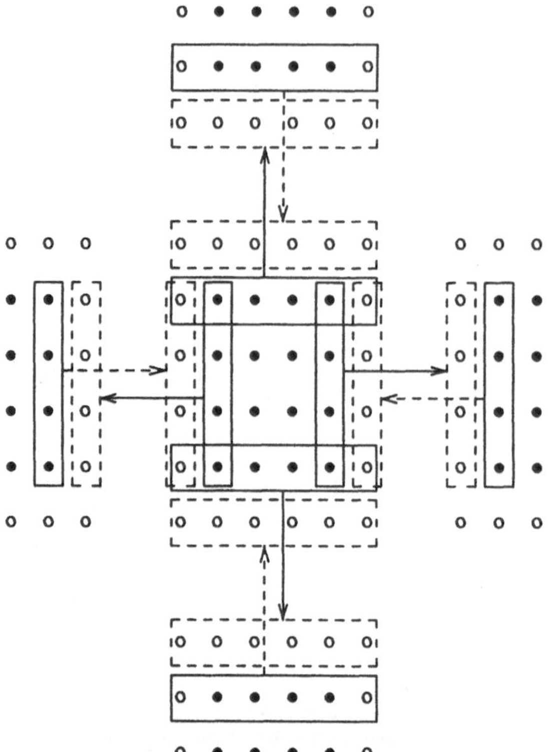

Figure 5.2. The 5-point flush operation

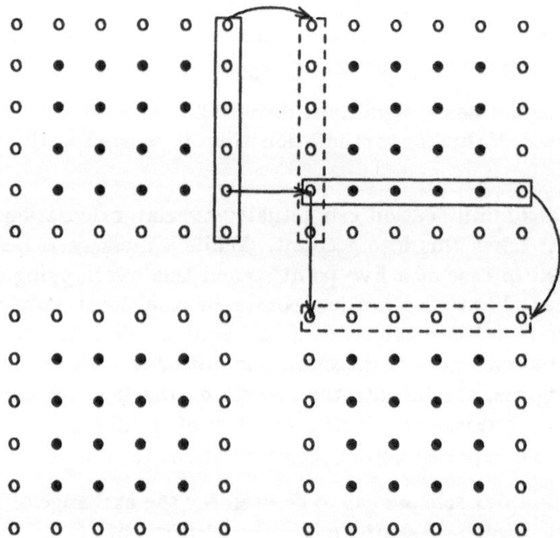

Figure 5.3. 9-point flush: path of an updated corner gridpoint

The flush operation arises in many implementations of problems that are defined on 2D- or 3D-grids. It is one of the necessary basic modules in any library of communication routines[9].

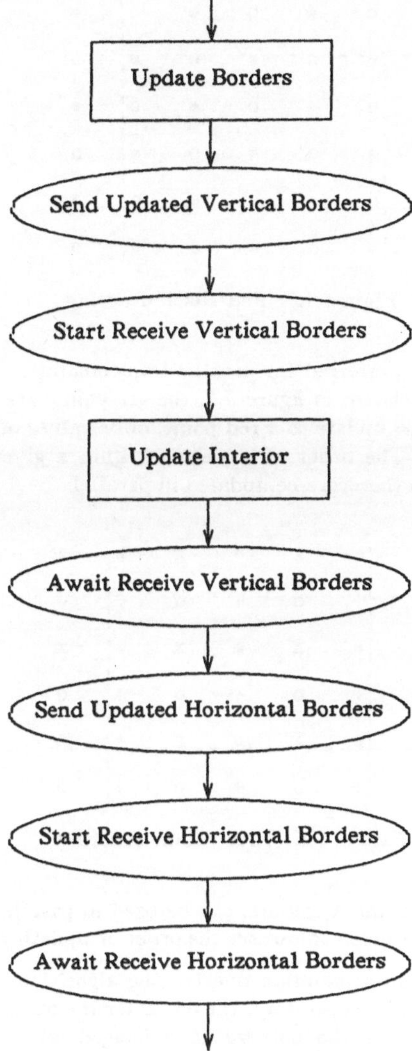

Figure 5.4. 9-point strategy

C. The Gauss-Seidel method

The Gauss-Seidel iteration consists of applying the scheme

$$(D + L)\, x^{(k+1)} = b - U\, x^{(k)} \tag{5.3}$$

repeatedly to an arbitrary starting solution. The method is different from the Jacobi algorithm since each newly calculated value is immediately used in the subsequent calculations. The ordering of the unknowns will therefore influence the result.

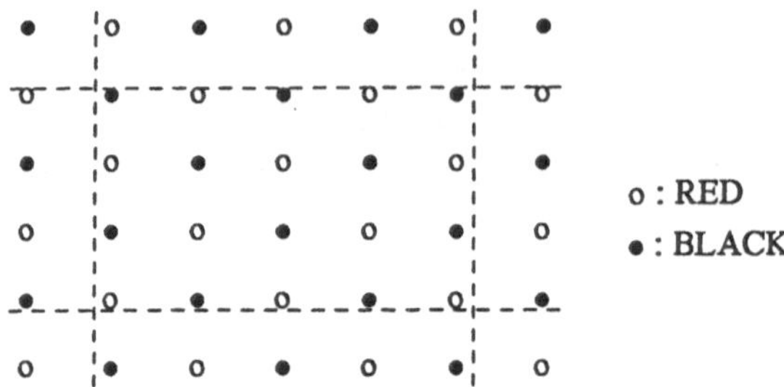

Figure 5.5. Red/Black ordering

The scheme which is usually preferred for parallel implementation is called the Red/Black or chequer board ordering. As shown in figure 5.5, the gridpoints are divided in a red and and a black subset. To calculate the update to a red point, only values of neighbouring black points are needed and vice versa. The order of updating within a given subset is arbitrary. All values of the same *colour* can therefore be updated in parallel.

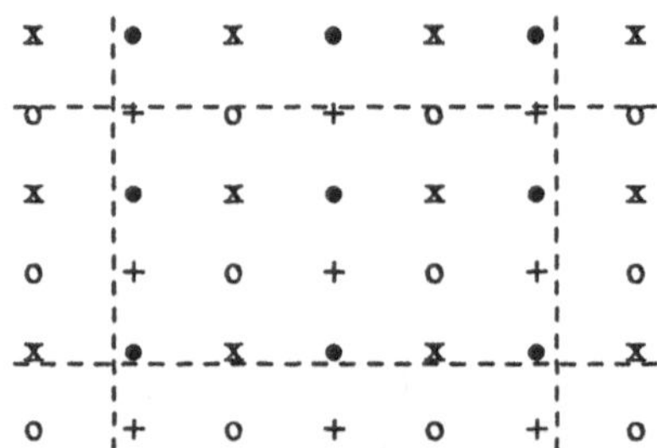

Figure 5.6. Four colour ordering

A similar colouring scheme, using 4 colours, can be used to parallelize a nine-point stencil, as shown in figure 5.6. Within a given colour set the order of updating is again arbitrary.

To insure communication-calculation overlap the algorithms of figures 5.1 and 5.4 are applied to each of the colour sets separately, the one after the other. The messages are shorter, since only a coloured subset of the borders is exchanged with the neighbours. The total number of exchanges per iteration however increases. As the message start-up time is the dominant factor, total communication time will increase for multicolour strategies.

D. The Successive Overrelaxation method

The Gauss-Seidel method is usually accelerated, leading to the SOR iteration:

$$(D + \omega L)\, x^{(k+1)} = \omega b + ((1-\omega)D - \omega U)\, x^{(k)} \tag{5.4}$$

with ω a relaxation parameter which can be chosen in such a way that the spectral radius of the iteration matrix is minimized. The equation for the i'th unknown x_i can be written as

$$x_i^{(k+1)} = \omega \left(b_i - \sum_{j=1}^{i-1} a_{ij} x_j^{(k+1)} - \sum_{j=i+1}^{n} a_{ij} x_j^{(k)} \right) / a_{ii} + (1-\omega) x_i^{(k)} \tag{5.5}$$

or

$$x_i^{(k+1)} = \omega\, x_i^{GS} + (1-\omega) x_i^{(k)}$$

with x_i^{GS} being the Gauss-Seidel update for $x_i^{(k)}$. The parallel characteristics of the SOR-method will therefore be similar to those of the Gauss-Seidel method.

E. The preconditioned conjugate gradient method

The conjugate gradient procedure for the solution of $A\,x = b$ where A is a positive definite matrix, is perhaps the most popular method for iteratively solving systems of linear equations. The convergence rate is slow for large problems. One way to improve this slow convergence is to precondition A.

The method consists of finding a nonsingular matrix C in such a way that $\hat{A} = C^{-1} A\, C^{-T}$ has an improved condition number[6]. The conjugate gradient algorithm can then be applied to the transformed system $\hat{A}\,\hat{x} = \hat{b}$, with $\hat{x} = C\,x$ and $\hat{b} = C^{-1}\,b$.

Another approach is to introduce a symmetric positive definite matrix $M = C\,C^T$ and to rewrite the standard algorithm directly in terms of M without explicitly forming C. The matrix M is called the *preconditioner*. M should be chosen in such a way that it is a good approximation to A and that the solution for any system $M\,z = r$ can be calculated efficiently.

One can then derive the following preconditioned conjugate gradient algorithm:

procedure *pcg*
  $r_0 = b - A\,x_0$
  solve $M\,z_0 = r_0$
  $p_0 = z_0$
  $\gamma_0 = \langle z_0, r_0 \rangle$
  for $k = 1, 2, 3, \cdots$
    $h_{k-1} = A\,p_{k-1}$
    $\alpha_{k-1} = \gamma_{k-1} / \langle p_{k-1}, h_{k-1} \rangle$
    $x_k = x_{k-1} + \alpha_{k-1}\,p_{k-1}$
    $r_k = r_{k-1} - \alpha_{k-1}\,h_{k-1}$
    if converged **then** terminate the loop
    solve $M\,z_k = r_k$
    $\gamma_k = \langle z_k, r_k \rangle$
    $\beta_k = \gamma_k / \gamma_{k-1}$
    $p_k = z_k + \beta_k\,p_{k-1}$
  endloop

The grid operations of the PCG algorithm can be subdivided into 4 classes:

— *preconditioner calculation* ($M\,z_k = r_k$):
Several efficient parallel preconditioners have been developed[14], with different communication requirements. The actual choice of a suitable preconditioner is largely problem dependent and still a very active area of research.
In the current status of this project only the very simple diagonal or 1-step Jacobi preconditioner has been implemented. It is perfectly parallel, since no communication is necessary.

— *matrix vector multiplication* (e.g. $h_{k-1} = A\,p_{k-1}$):
Each processor stores the equations associated with the gridpoints laying in its interior domain. The matrix vector product is therefore readily calculated without any communication. One flush operation may be needed afterwards to assign correct values to the gridpoints in the overlap region.

— *vector plus scalar times vector* (e.g. $x_k = x_{k-1} + a_{k-1}\,p_{k-1}$)
No exchange of information is necessary. All values can be calculated in parallel. If the values in the overlap region are needed, they can be calculated. They needn't be communicated.

— *scalar product* (e.g. $\gamma = \langle z_k, r_k \rangle$):
The scalar product is typically computed in two stages. First, part of the scalar product is calculated in each processor. These intermediate results are subsequently combined to form the global scalar product. On hypercube machines, the combination is usually performed in a tree-wise manner, see figure 5.7.

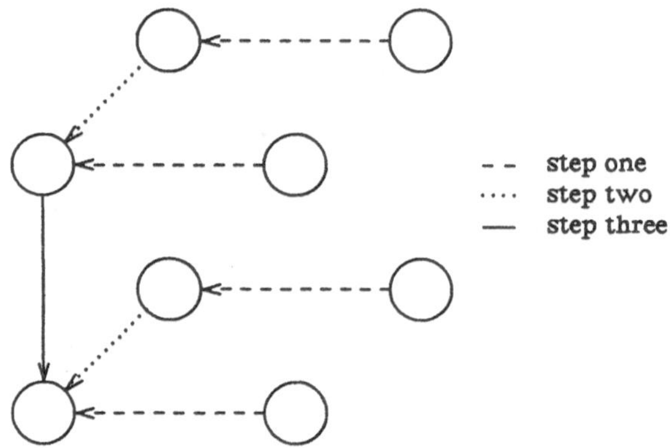

Figure 5.7. Communication strategy of the global combine operation

The global result is formed in processor zero, and broadcasted to all nodes. A faster combination that uses only half of the number of communication steps of the previous algorithm is explained elsewhere[4].

Each of the operations of the conjugate gradient method thus parallelizes nicely.

F. The multigrid method

1. Introduction

The convergence of the Gauss-Seidel iteration is typically fast in the first few steps and then slows down. It quickly reduces the high frequency components of the error, but not the low frequency ones. If the error is transferred onto a coarser grid, the lower frequencies become the higher frequencies and therefore can be annihilated more rapidly than on the fine grid.

This principle is taken advantage of in the multigrid method. The method differs from the other iterative techniques in that it uses a set of nested grids, with the finest one corresponding to the one on which the solution is desired. A detailed discussion can be found in the standard reference texts[7,8]. In the next section we shall briefly review the basic multigrid idea.

2. The multigrid algorithm

Let G^i, $i = 0, 1, \cdots, l$ be the hierarchy of grids with G^l the finest grid and G^0 the coarsest grid. The linear elliptic differential operator is discretized on each grid, giving the discrete operator L^i. Let u_a^i be an approximation to the solution of the equation

$$L^i u^i = f^i \quad \text{on } G^i , \tag{5.6}$$

obtained e.g. by applying a few Gauss-Seidel iterations to an initial approximation for u^i. For a suitably chosen relaxation the error $e^i = u^i - u_a^i$ will be a smooth function on G^i. The relaxation is therefore called a *smoothing* operation. The error e^i satisfies the defect equation

$$L^i e^i = d^i \qquad \text{with } d^i = f^i - L^i u_a^i. \tag{5.7}$$

In the multigrid method the defect equation is projected onto the coarser grid G^{i-1} using a *restriction* operator I_i^{i-1}

$$L^{i-1}\, u^{i-1} = I_i^{i-1} d^i \tag{5.8}$$

This equation is solved for u^{i-1}, using any method applicable on G^{i-1}. The result is interpolated back to the fine grid G^i, using an interpolation or *prolongation* operator I_{i-1}^i. The approximate solution u_a^i is then corrected following

$$u_a^i := u_a^i + I_{i-1}^i\, u^{i-1} \tag{5.9}$$

Equation 5.8 is in nature similar to the problem we started from. It can be solved by recursively applying the same solution method. This leads to the following algorithm.

procedure *mgm* (L^i, f^i, u^i)

if G^i *is the coarsest grid, i.e.* $i = 0$ **then**
  solve $L^i\, u^i = f^i$ *exactly*
else
  - *perform* ν_1 *smoothing operations*
  - *compute the defect:*
    $d^i := f^i - L^i\, u^i$
  - *project the defect on* G^{i-1}:
    $f^{i-1} := I_i^{i-1}\, d^i$
  - *solve the coarse grid problem* $L^{i-1}\, u^{i-1} = f^{i-1}$ *recursively by* γ_i *iterations of mgm, i.e.*
    repeat γ_i **times** *mgm*$(L^{i-1}, f^{i-1}, u^{i-1})$, *starting with* $u^{i-1} = 0$.
  - *interpolate the correction from* G^{i-1} *back to* G^i *and correct* u^i:
    $u^i := u^i + I_{i-1}^i\, u^{i-1}$
  - *perform* ν_2 *smoothing operations*

endif

The algorithm is normally combined with the idea of nested iteration. The initial approximation to the problem on G^i is then derived from the solution obtained on G^{i-1}. This leads to the *full multigrid method*.

procedure *fmg*

solve the coarse grid problem $L^0\, u^0 = f^0$
for $i = 1$ **to** l **do**
  interpolate the solution on G^{i-1} *to* G^i
    $u^i = I_{i-1}^i\, u^{i-1}$
  solve the problem on grid G^i
    repeat δ **times** *mgm*(L^i, f^i, u^i)
endloop

The algorithm is completely defined by specifying the grid sequence G^i, the discretized operator L^i, the inter-grid transfer operations I_{i-1}^i and I_i^{i-1}, the nature of the smoothing relaxations, and by assigning a value to the constants ν_1, ν_2, γ_i and δ. So-called V- and W-multigrid-cycles are obtained with the values 1 and 2 for γ_i. Another choice of γ_i leads to what is known as an F-cycle. In figure 5.8 we depicted the order in which the different gridlevels are visited for each of the three important cycle types.

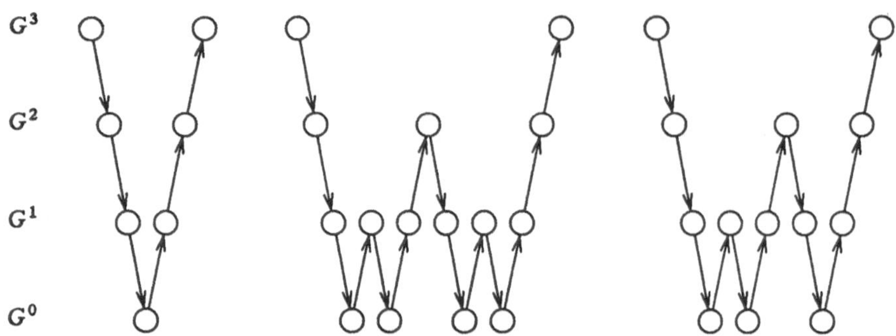

Figure 5.8. V-, W- and F-cycle

The correct choice of the multigrid modules is of crucial importance to the convergence behaviour and is far from trivial[8]. If all parameters have been chosen optimally it can be proven that the rate of convergence is independent of the size of the problem. The discretization error is obtained in $O(n)$ operations, with n the number of unknowns.

In the current implementation of our library, standard coarsening is used. Mesh sizes are halved in both directions for each level of grid refinement. The discrete operator L^i is obtained by discretizing the elliptic differential operator on G^i. A Red-Black Gauss-Seidel method is used as a smoothing relaxation.

3. Parallelism

A *fixed partitioning scheme* is used to allocate subdomains to processors. Each processor is assigned a fixed part of the rectangle Ω, the domain of the *continuous* problem, and is responsible for doing the computations on all gridpoints belonging to that subdomain. With very coarse grids, some processors may have no gridpoints in their subdomain. In a multigrid application a processor can therefore be active on some (fine) grids, and inactive on other (coarse) grids.

Each of the constituent multigrid operations (smoothing, defect calculation, restriction and prolongation) is basically a local operation and therefore parallelizes nicely. The communication requirements are however slightly different. In the following discussion we assume that all grids that are acted upon by the modules are consistent. The values of the gridpoints in the overlap region are correct.

a. The smoothing operation:

The communication aspects of the five-point and nine-point Gauss-Seidel relaxation have been discussed before. In the latter case all values in the overlap region are correct after the relaxation. In the five-point case the corner values of the overlap region are not used in the calculations and are therefore never updated. If a consistent grid is needed in subsequent calculations, a flush operation should be executed.

b. The defect calculation:

No communication is required to calculate the defect in the gridpoints interior to a processors subdomain. Whether the values in the overlap region are needed, depends on the operation that follows, i.e. the nature of the restriction. One flush operation may therefore be needed.

c. The restriction:

The calculation of the restriction is an inter-grid operation. It transfers information from a fine grid G^i to a coarse grid G^{i-1}.

Several restriction formulae can be used. Let (K,L) and (k,l) be coarse grid and fine grid coordinate pairs of a same physical point in grid G^{i-1} and G^i. Typical restriction formulae are given by

- injection: $u_{KL} = u_{kl}$

- half-injection: $u_{KL} = \frac{1}{2} u_{kl}$

- full-weighting: $u_{KL} = \frac{1}{16} \begin{bmatrix} 1 & 2 & 1 \\ 2 & 4 & 2 \\ 1 & 2 & 1 \end{bmatrix} u_{kl}$

The last formula is modified near the boundaries where a mixed condition holds[12].

All interior points can be calculated in parallel. The values in the overlap area need to be communicated. When the processor is inactive on G^{i-1} no calculations have to be done. The processor is then able to continue with other work or waits until the other processors have finished their activities on G^{i-1} and the coarser grids.

d. The prolongation:

Bilinear interpolation is used to prolongate gridpoint values from G^{i-1} to G^i.

The extended subdomain associated with the fine grid is completely contained in the extended region of the coarse grid. All gridpoint values, the interior as well as the values in the overlap region, can therefore be calculated without any need for message passing.

A substantial problem arises when a processor is active on the fine grid, but inactive on the coarse grid. A typical situation is shown in figure 5.9. The coarse grid contains 2 gridlines in x- and y-direction. Sixteen processors are mapped onto the solution domain. Only 4 processors are active on the coarse grid, yet all the processors are active on the fine grid.

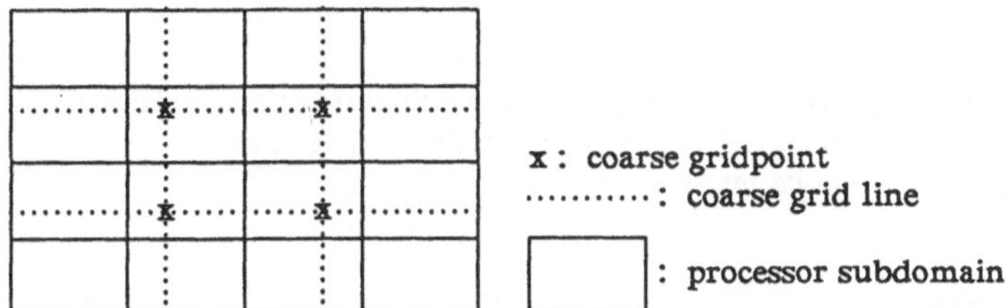

Figure 5.9. Idle processors on the coarse grid

The inner four processors, which are active on both grids, can easily calculate the prolongation for each gridpoint in their extended subdomain. The inactive processors cannot calculate the interpolated values at the fine grid points since they lack any information on the coarse grid. They were idle when computations proceeded on the coarse grid. To be able to resume computation, they need to be *revived*. The fine grid values are to be sent to them.

The reviving of idle processors happens in two stages as shown in figure 5.10. In the first step the necessary values are sent to the nodes horizontally and vertically. In the second step the information is sent to the processors in the diagonal directions.

A weakness of the parallel implementation of standard multigrid is that many processors are left inactive during coarse grid calculations. This problem is often referred to as the *processor-idling problem*. A second problem is due to the small number of unknowns each processor contains on coarse grids. It is also due to the complex procedure to revive an inactive processor in the prolongation procedure. The *high communication to calculation ratio* will seriously degrade the speed-up obtainable on the coarse grids.

The relative frequency by which the coarse grids are visited will thus be of crucial importance to the obtainable efficiency. On a sequential machine the W-cycle is usually preferred. As shown in figure 5.8 the coarse grids are visited very frequently. On a parallel machine V- and F-cycles may be faster.

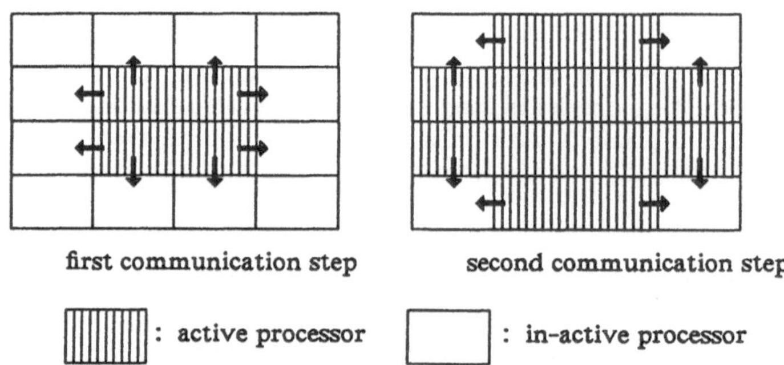

Figure 5.10. The process of reviving idle processors

A solution to the processor-idling problem requires innovative multigrid methods[5]. The second problem can be alleviated by applying different partitioning strategies. The strategy based on *grid-agglomeration* seems to be particularly successful[10].

VI. TIMING RESULTS

A. Introduction

Extensive timing results for several test problems can be found in the companion report[20]. In this section we present timing results for a simple Helmholtz problem taken from the ELLPACK pde problem set (ref 17, p.449):

$$\frac{\partial^2 u}{\partial x^2} + \frac{\partial^2 u}{\partial y^2} - 10\,u = f \quad \text{on} \quad \Omega = [0,1] \times [0,1] \tag{6.1}$$

with Dirichlet boundary conditions chosen so that the solution is given by

$$u(x,y) = \cos(\pi y)\,\sin(\pi(x-y)). \tag{6.2}$$

A contour plot of the solution is shown in figure 6.1. This particular solution was calculated on a 47 by 47 grid using 8 processors. The discretization points and the division of the domain into subdomains are also shown.

B. Parallel efficiency

The computational efficiency ϵ of our implementation of the Jacobi method for different processor configurations is shown in figures 6.2 and 6.3.

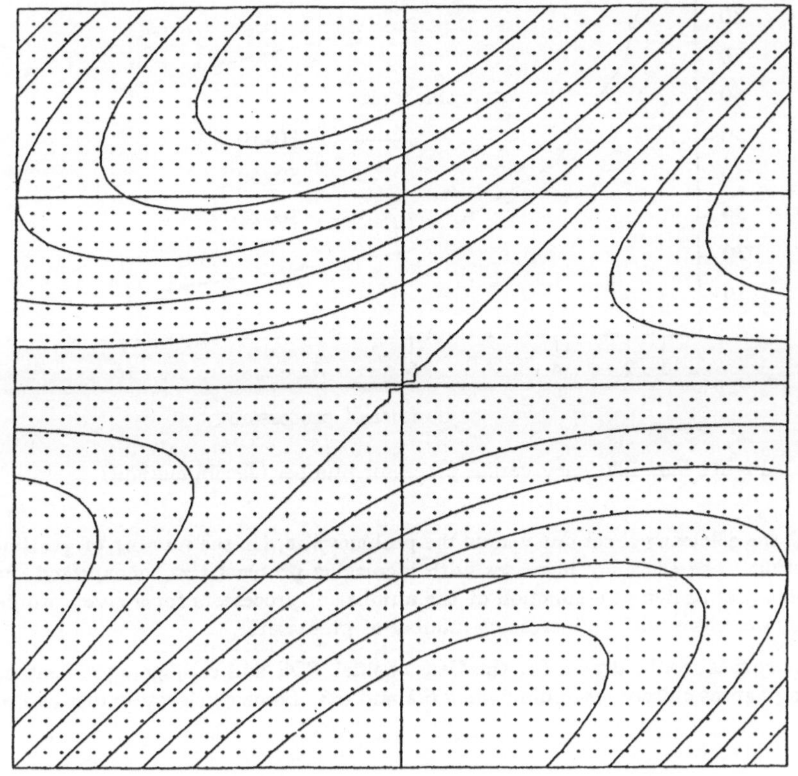

Figure 6.1. Contourplot of solution (6.2) calculated on a 2 by 4 processorgrid

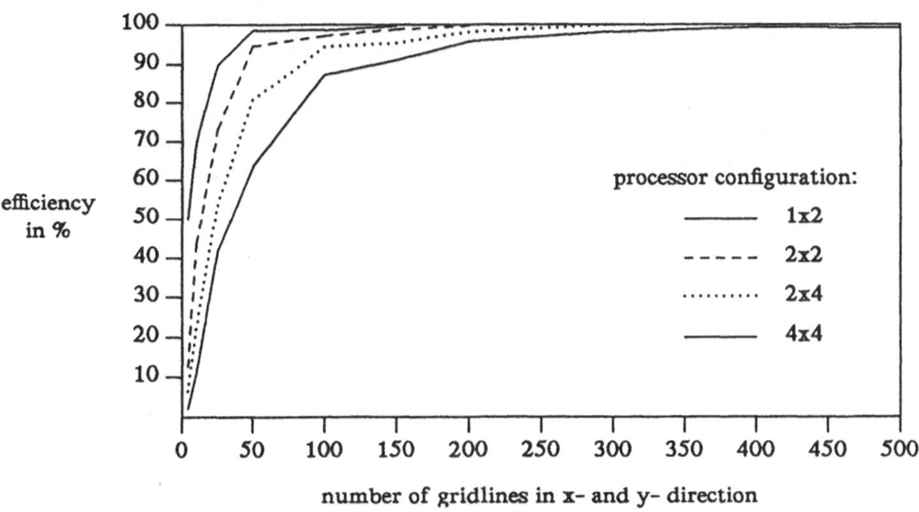

Figure 6.2. Efficiency versus problem size (Jacobi method)

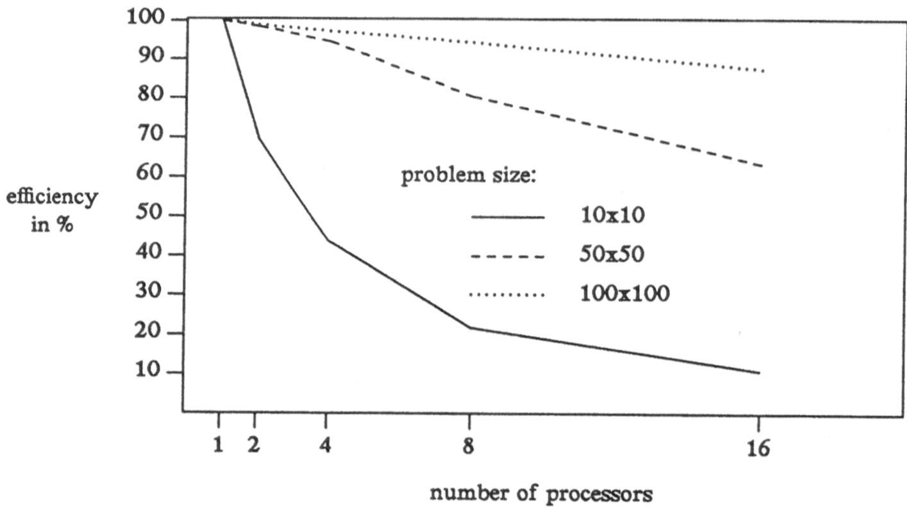

Figure 6.3. Efficiency versus processor number (Jacobi method)

In figure 6.2 the efficiency is set out versus the problem size, i.e. the number of gridlines in x- and y-direction. The efficiency increases with increasing problem size, as expected. In figure 6.3 the efficiency is drawn as a function of the number of processors for a given problem size. The efficiency decreases with an increasing number of processors.

Similar curves can be drawn for the Gauss-Seidel, SOR and conjugate gradient methods. The resulting figures closely resemble figures 6.2 and 6.3. Some efficiency values can be found in table 6.1.

Table 6.1: 5 point stencil efficiency, 4 by 4 processor configuration

problem size:	10	25	50	100	200	300
ϵ_{Jacobi}	10.9	42	63.5	87.4	95.7	98.2
ϵ_{SOR}	4.2	20.4	50.3	77.7	93.4	96.9
ϵ_{cg}	7.6	31.0	56.9	84.6	95.3	98

The parallel efficiency of the multigrid method depends very much on the kind of multigrid cycle. In figure 6.4 we plotted the efficiency versus the problem size for a *V(1,1)*-cycle, i.e. a V-cycle with one pre-smoothing and one post-smoothing relaxation. Half-injection was used as the restriction formula. The coarsest grid contained only one point. In the given multigrid method the computational complexity reduces by a factor of four for each transition to a coarser grid. In a V-cycle each grid is visited only once. The inefficiency of the coarse grid iterations therefore only slightly reduces the overall parallel efficiency.

A similar curve is drawn in figure 6.5 fo a $W(1,1)$-cycle. The coarsest grid G^0 is now visited 2^l times for each W-cycle on the fine grid G^l. The inefficient coarse grid operations now largely deteriorate the global performance. From the reasons given at the end of section 5 it can be understood that this will always happen for small problems and/or large processor numbers.

For each F-cycle on G^l the coarse grid G^0 is visited l times. The efficiencies of the F-cycle will therefore be between those for the V- and W-cycles.

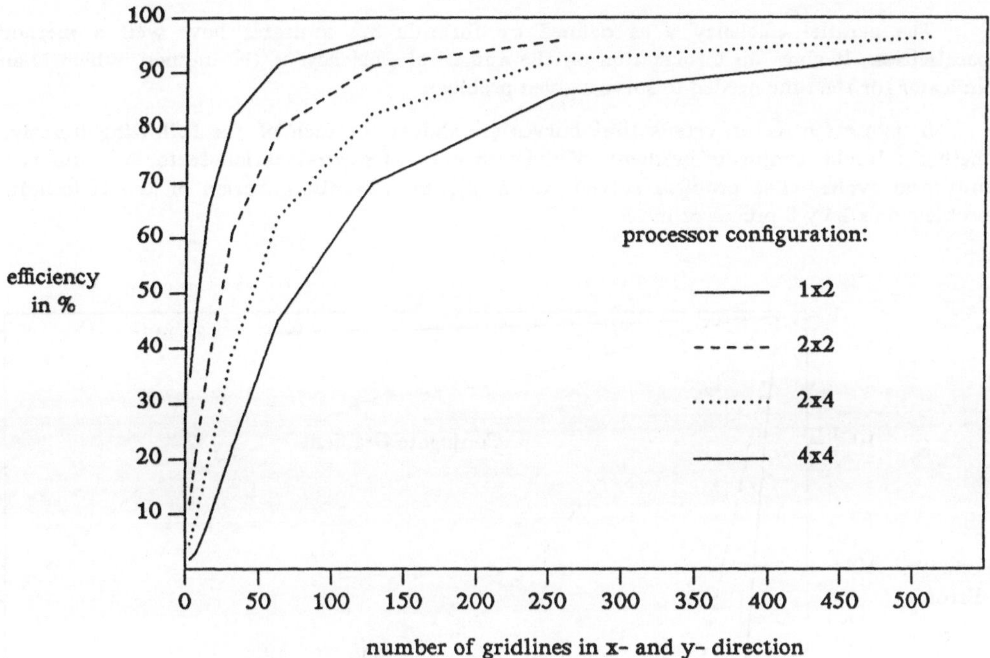

Figure 6.4. Efficiency of a V(1,1)-cycle, 5-point stencil

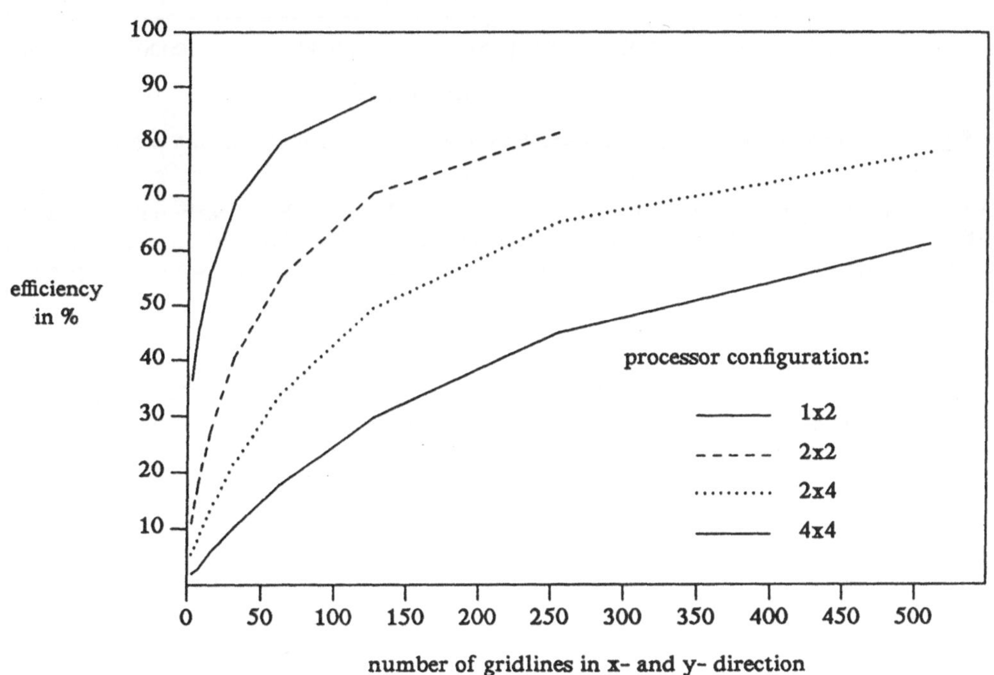

Figure 6.5. Efficiency of a W(1,1)-cycle, 5-point stencil

C. Numerical efficiency versus parallel efficiency

The parallel efficiency ϵ as defined by formula 3.2 indicates how well a method parallelizes. It gives no information on the numerical efficiency of the method, which is an indicator for the time needed to solve a given problem.

In figure 6.6 error versus time curves are shown for each of the following iterative methods: Jacobi, conjugate gradient, SOR (with optimal overrelaxation factor ω), and two multigrid cycles. The problem solved was a 127 by 127 discretization of the Helmholtz problem on a 2 by 8 processor mesh.

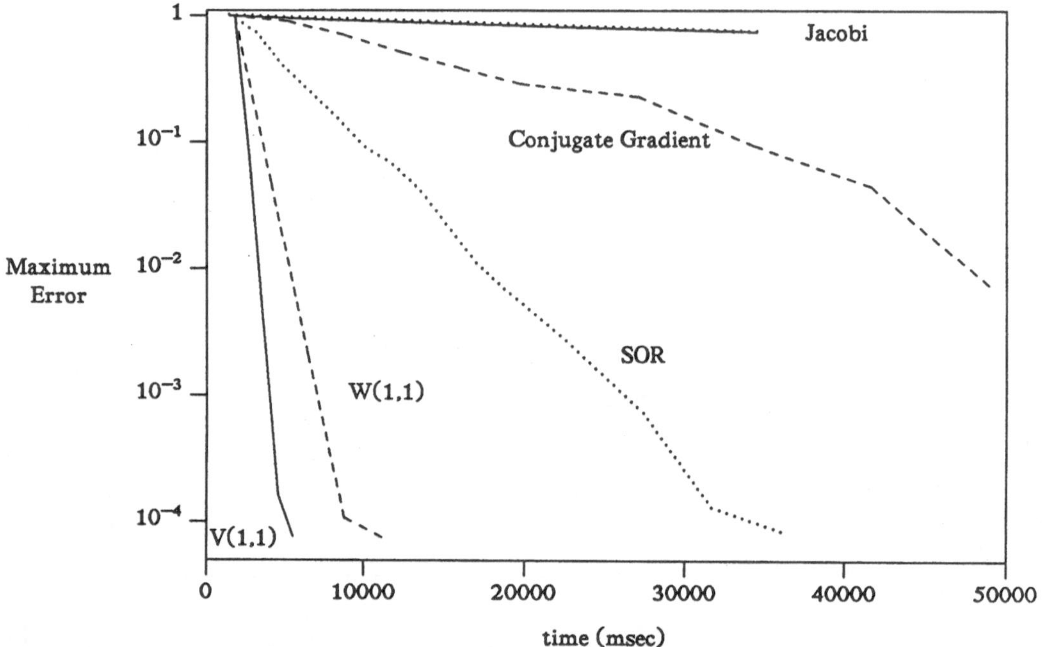

Figure 6.6. Error-time curves for the solution of a 127 by 127 problem on 8 processors

Although the multigrid methods have a lower parallel efficiency, they clearly outperform the other iterative methods. The W-cycle, which is normally competitive with or even preferred to the V-cycle on a sequential machine, shows to be the slower of the two on the parallel machine.

Acknowledgement

We would like to acknowledge Kurt Wayenberg for his programming help in the early stages of the project, Luc Bomans for his continuous system support, Dirk Roose and Nele Geurden for carefully reading this manuscript.

[1] L. Bomans and D. Roose, *Timing results on the Intel iPSC/2*, report TW, Katholieke Universiteit Leuven, Belgium, (in preparation).

[2] A. Brandt, *Guide to multigrid development* in *Multigrid Methods*, (W. Hackbusch, U. Trottenberg, eds.), Lecture Notes in Mathematics 960, Springer Verlag, Berlin, 1982.

[3] T.F. Chan, Y. Saad and M. Schultz, *Solving elliptic partial differential equations on hypercubes* in *Hypercube Multiprocessors 1986*, SIAM, Philadelphia, 1986.

[4] G.C. Fox, M.A. Johnson, G.A. Lyzenga, S.W. Otto, J.K. Salmon and D.W. Walker, *Solving Problems on Concurrent Processors*, Prentice Hall, 1988.

[5] P.O. Frederickson and O. McBryan, *Parallel Superconvergent Multigrid*, presented to 3rd Multigrid Conference, Copper Mountain, April 8 1987.

[6] G.H. Golub and C.F. Van Loan, *Matrix Computations*, John Hopkins Press Baltimore, 1984.

[7] W. Hackbush, *Multi-grid methods and Applications*, Springer Series in Comp. Math. 4, Springer-Verlag, Berlin, 1985.

[8] W. Hackbush and U. Trottenberg (eds.), *Multigrid Methods*, Lecture notes in Mathematics 960, Springer-Verlag, Berlin, 1982.

[9] R. Hempel, *The SUPRENUM Communications Subroutine Library for Grid-oriented Problems*, report ANL-87-23, Argonne National Laboratory, Illinois, 1987.

[10] R. Hempel and A. Schuller, *Experiments with Parallel Multigrid Algorithms, Using the SUPRENUM Communications Subroutine Library*, preliminary GMD report, St-Augustin, West Germany, 1988.

[11] *iPSC/2 Users Guide*, Intel Scientific Computers, Beaverton, 1987.

[12] W. Joppich, *A Multigrid Method for Solving the Nonlinear Diffusion Equation on a Time-Dependent Domain Using Rectangular Grids in Cartesian Coordinates*, Arbeitspapiere der GMD 259, St-Augustin, West Germany, 1987.

[13] D. Kamowitz, *SOR and MGR[ν] experiments on the Chrystal multicomputer*, Parallel Computing 4, pp. 117-142, 1987.

[14] D.E. Keyes and W.D. Gropp, *A comparison of Domain Decomposition Techniques for Elliptic Partial Differential Equations and their Parallel Implementation*, Research Report YALEU/DCS/RR-448, 1985.

[15] O. McBryan and E. Van de Velde, *Hypercube Algorithms and Implementations*, SIAM J. Sci Stat. Comput. 8, pp 227-287, 1987.

[16] J. Ortega and R. Voigt, *Partial Differential Equations on vector and parallel computers*, SIAM Rev. vol. 27, pp. 213-240, 1985.

[17] J.R. Rice and R.F. Boisvert, *Solving Elliptic Problems Using ELLPACK*, Springer Series in Computational Mathematics 2, Springer Verlag, New York, 1985.

[18] C.L. Seitz, *The Cosmic Cube*, CACM, pp. 22-33, 1985.

[19] C. Thole, *Experiments with Multigrid on the Caltech Hypercube*, GMD-Studien Nr. 103, St-Augustin, West Germany, 1985.

[20] S. Vandewalle, J. De Keyser and R. Piessens, *The Numerical Solution of Elliptic Partial Differential Equations on a Hypercube Multiprocessor*, report TW, Katholieke Universiteit Leuven, in preparation.

II. SUPERCOMPUTERS LANGUAGE AND ALGORITHMS

DESIGN OF NUMERICAL ALGORITHMS FOR SUPERCOMPUTERS

D.J. Evans

Department of Computer Studies, Loughborough University of Technology, Loughborough, Leicestershire, U.K.

1. INTRODUCTION

Parallelism arises at many different levels within a complex problem which if exposed can be efficiently exploited By incorporating software tools in the parallel system to measure the performance we are able to restructure our algorithms or component parts of them into parallel form to run more efficiently on parallel computers.

In particular, we have studied the various standard techniques of achieving parallelism, i.e. vectorization, problem partitioning and divide and conquer strategies as well as exploiting the use of implicit parallelism in various numerical algorithms. In addition, some new parallel algorithms have been introduced.

The principles we have learnt from this study have been extended to algorithms on parallel systems, i.e. the DAP and CRAY.

2. NUMERICAL PARALLEL ALGORITHMS

The main aims of our work have been as follows:

1. To discover and design alternative solution methods which offer parallelism in one form or another.
2. To study the suitability of each parallel scheme for implementation on different parallel computer systems.
3. To obtain the performance analysis of the implemented procedures.

The primary feature that distinguishes parallel algorithms and systems from the more usual uniprocessor situation is that parallelism entails the use of facilities or resources not present in sequential solutions, i.e. namely:

i) multiple processors
ii) data communication
iii) synchronisation to determine the state of related processors (a special type of communication).

The introduction of these factors into the computation can make significant changes in the algorithm design.

2.1. Algorithm Structure

Algorithms can contain parallelism at different levels which may:

1. be apparent in the high level problem specification
2. arise from the method of solution
3. arise in the details of the solution.

As an example consider Numerical Quadrature: its specification is the sum of several independent function evaluations. Methods of solution are often based on successively splitting the domain of integration into smaller domains to each of which the quadrature specification is applied independently. Finally, the detailed examination of the integrand can also reveal independent computations as well as forms of implicit parallelism.

To evaluate $\int_a^b f(x)dx$, where $f(x)$ is continuous in (a,b), then by partitioning the range (a,b) into n subintervals of size, i.e. $(a,x_1),(x_1,x_2)\ldots(x_i,x_{i+1})\ldots(x_n,b)$, the integral can be written in the form,

$$A = \int_a^{x_1} f(x)dx + \int_{x_1}^{x_2} f(x)dx + \ldots\int_{x_i}^{x_{i+1}} f(x)dx + \ldots \int_{x_n}^{b} f(x)dx \ .$$

By use of the simple trapezoidal quadrature rule we can determine each of the integrals in the form, i.e.,

$$A_1 = \int_a^{x_1} f(x)dx = \frac{h}{2}\ [f(a)+f(x_1)\ ] + \frac{h^2}{12}\ f''(\xi),\ a < \xi < x_1 .$$

Finally, the numerical quadrature can be completed in parallel by use of the fan-in algorithm which for n = 8 is,

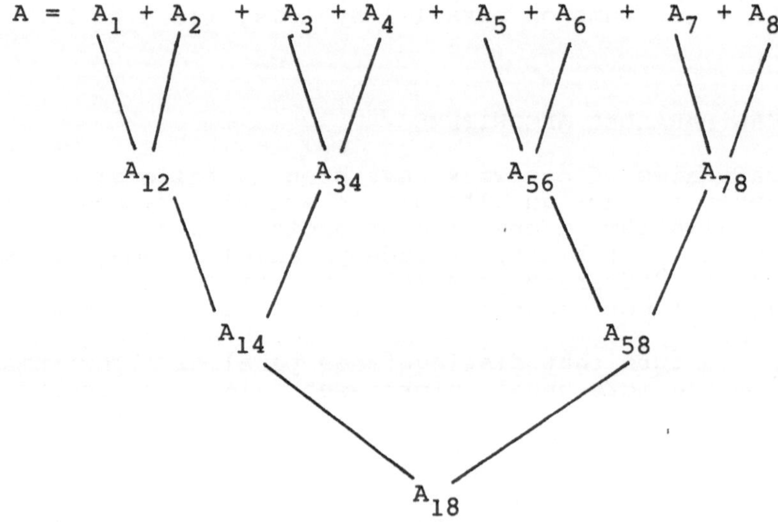

which can be completed in logN(=3) steps.

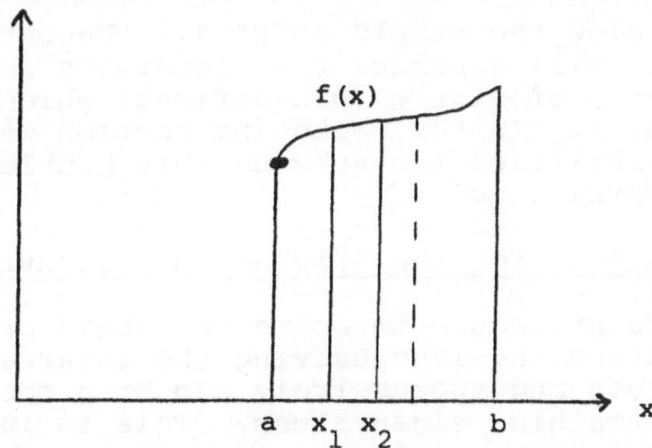

Generally, for the design of effective parallel algorithms one has to match the resources demanded by an algorithm and the resources available on a given parallel architecture. This involves:

a) The ability to express the solution methods for the problem in hand in terms of independent (parallel) processes. The nature of the operations required by each or all of the processes has to be matched with the processing capabilities of the processing elements within a given system. On the other hand, the number of available processes and the allocation cost of processes to processors should be a determining factor on how far a solution method can be expressed in terms of parallel processes.
b) The evaluation of the communication/computational bandwidth required by the algorithm and that offered by the various parallel computer systems to determine (and minimise if possible) the overhead cost associated with the required communication.

To illustrate the above points consider the following problem of evaluating the eigenvalue λ_1 of a symmetric tri-diagonal matrix A (b_i, c_i, b_{i+1}), $i = 1,2,\ldots,n$; $b_1 = b_{n+1} = 0$, i.e.

$$A = \begin{bmatrix} c_1 & b_2 & & & \\ b_2 & c_2 & b_3 & & O \\ & \ddots & \ddots & \ddots & \\ & O & \ddots & \ddots & b_n \\ & & & b_n & c_n \end{bmatrix}$$

The solution algorithm involves the repeated evaluation of the recurrence relation,

$$q_0 = 1, \; q_1 = c_1 - \lambda_1,$$
$$q_i = (c_i - \lambda_1) - b_i^2/q_{i-1}, \; i = 2,3,\ldots,n, \qquad (2.1)$$

for different values of λ_1. The number of negative q_i's, $i = 1,2,\ldots,n$ of the above sequences will indicate the number of eigenvalues below the sample point λ_1. Repeated application of this procedure will separate the eigenvalue spectrum into small sub-intervals of size ε (pre-defined) which contain 1 or more eigenvalues, λ. In the following section we shall briefly discuss the possibilities for solving this problem on alternative parallel systems.

2.2. Parallel Methods for the Tridiagonal Eigenvalue Problem

The standard procedure to solve the above problem on sequential computers involves halving the interval containing all the eigenvalues and successively choosing one of the two sub-intervals containing eigenvalues. This is in turn bisected into two further sub-intervals and the process is continued until the eigenvalues are separated to a predefined accuracy. The method of parallel bisection (Barlow and Evans, [1]) uses the principle on all the previously determined non-empty sub-intervals. A major disadvantage with this scheme is that the number of parallel processes available at the initial stages of the algorithm is limited to 1 at the 1st iteration, 2 in the 2nd iteration, 4 in the 3rd iteration and so on. Therefore, the potential speed up obtainable from this parallel version is dependent upon the number of non-empty intervals available at each stage and is bounded by N, the maximum number of eigenvalues.

However, in the multisection procedure, each sub-interval is divided into m (instead of 2 as in the bisection procedure) sub-intervals for each of which the Sturm sequence is evaluated in parallel on each of the available processors simultaneously. This procedure can be extended further into a method called parallel multisection (Barlow et al. [2]) in which the above two methods are combined to obtain greater efficiency. In this method given p processors and m domains then one allocates $\ell = p/m$ processors per domain. The speed-up of the resultant method lies between m and $m \log_2(\ell+1)$.

In all of the above methods the recurrence relation is evaluated sequentially and parallelism is generated through the simultaneous evaluation of several sequences for different sample points. However, additional parallelism through the reformulation of the Sturm sequence itself can also be achieved as the following two algorithms will show.

The method of recursive doubling (Lambiotte, [3]), for example, re-defines the Sturm sequence as:

$$p_0 = 1,\ p_1 = c_1 - \lambda_1\ ,$$
$$p_i = (c_i - \lambda_1)p_{i-1} - b_i^2 p_{i-2}\ , \quad i = 2,3,\ldots \qquad (2.2)$$

which can be expressed in the form,

$$\begin{bmatrix} p_i \\ p_{i-1} \end{bmatrix} = \begin{bmatrix} \prod_{j=1}^{i} S_j \end{bmatrix} \begin{bmatrix} p_1 \\ p_0 \end{bmatrix}\ , \quad i = 2,3,\ldots,n \qquad (2.3)$$

where,

$$S_j = \begin{bmatrix} c_j-\lambda & -b_j^2 \\ 1 & 0 \end{bmatrix} \quad (2.4)$$

It can be easily seen that the sequences p_i and q_i, $i = 1,2,...,$ n are related through the formula,

$$q_i = \frac{p_i}{p_{i-1}} \quad (2.5)$$

The parallelism in the above procedure is obtained by evaluating the (2*2) matrix multiplication terms $\prod_{j=1}^{i} S_j$, $i = 2,3,...,n$ in the following manner, i.e. for n = 8

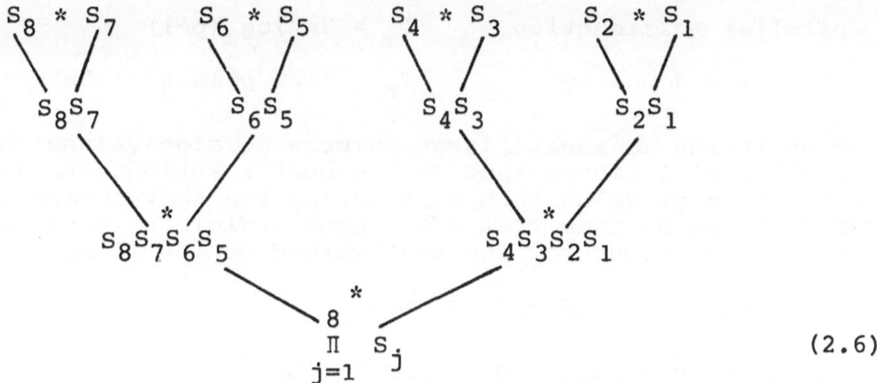

(2.6)

It can be seen from (2.3) that all the terms are built up of products of the S_j matrices. The system can be solved in $\log_2(n)$ sequential stages, by the fan-in network, each stage consisting of between n/2 and n parallel processes each of which is a multiplication of (2*2) matrices. Thus, stage 1 forms all products $(S_j.S_{j-1})$ for $j = 2,3,...,n$, stage 2 combines all the results to give the products $(S_j.S_{j-1}).(S_{j-2}.S_{j-3})$ for $j = 4,5,...,n$ and so on.

Thus, the parallel evaluation of the Sturm sequence can be completed in a time

$$T_r = (12n/p)\log_2(n) + 3n/p$$

using p processors and taking into account the cost of counting the sign changes in the p_i sequence.

2.3. Analysis of the Alternative Solution Methods

To be effective, a parallel solution must not only be accurate and fast relative to the same algorithm run on a uni-processor but it must also be efficient relative to the best sequential algorithm for that problem. Therefore, there is little use if the sequential starting point is inefficient or the cost of the reformulation of a sequential procedure into a parallel version is too high. For example, the cost of the evaluation of the Sturm sequence sequentially is 3n floating point operations (flops) where n is the size of the system. The recursive doubling procedure for evaluating the same sequence costs 12n flops. Therefore, it can be seen that the introduction of parallelism within the evaluation of the recurrence relation has produced a procedure which is immediately more expensive by a factor of 4. Thus, since the maximum speed up of the recursive doubling procedure is only $\log_2 p$ the actual speed up compared with the sequential bisection is $\log_2 p/4$ if the processors on the parallel system have equal computing power to that of the sequential system.

To compare the performance of the above alternative solution methods we analyse their individual computing times for evaluating a single eigenvalue as follows:

sequential bisection	$T = 3n$
parallel bisection (N>p)	$T_b = 3n/p$
parallel multisection	$T_m = 3n/\log_2(p+1)$
recursive doubling	$T_r = (12n/p)\log_2(n)+3n/p$

When trying to locate large numbers of eigenvalues (N>p) the parallel bisection method is the best solution. However, for a single eigenvalue then considering the above computational costs it can be seen that after some judicious reasoning the following conclusions for the best method can be made, i.e.

$$T_r < T_m \text{ if } p/\log_2 p > 4\log_2(n).$$

Thus, for n = 1024, for example, the recursive doubling strategy is better than multisection if the number of processors is greater than 512 which signifies that it is suitable for the DAP but not for the CRAY.

Another important criterion in designing effective parallel algorithms is the stability of the proposed solution methods. Unstable solutions are of little interest to the user and current sequential algorithms have had their stability properties thoroughly analysed. Therefore, any parallel solution that deviates from the calculation of a sequential solution must be thoroughly analysed for stability.

It can be easily seen that the computations involved in the parallel bisection and multisection algorithms follow that of the sequential bisection method. However, in the recursive doubling method the computation involves matrix multiplications instead of scalar operations as in the previous procedures. There still remains further work to be done, to verify the stability characteristics of these procedures. For example, in the recursive doubling procedure to produce results of identical accuracy to that obtained by the standard bisection method all the elements involved in the matrix operations have to be stored in double precision. This in turn will affect the performance of the algorithm on systems such as DAP where the performance of the system as a whole degrades as the word length increases (e.g. on the DAP, the timing for the multiplication of two numbers stored as R*4 is 274 μseconds whereas for R*8 it is 1066 μseconds.

Table 2.1. ICL-DAP and CRAY-1 timings for locating all the eigenvalues using combined multi-section and bisection.

SIZE	ICL-DAP		CRAY-1	
	Time (secs.)	Speedup*	Time (secs.)	Speedup***
64	0.24	4	0.028	3.8
256	1.15	12	0.27	5.5
1024	6.66	27	3.14	6.2
4096	65.15	>46**	49.26	6.8

*speedup calculated with respect to ICL 2980 (the DAP host)

**ICL 2980 version ran out of time

***compared to CRAY sequential solution

2.4. Results

Combined multi-section and bisection has been successfully implemented on 2 different types of parallel architectures. These are the ICL-DAP, an array processor and the CRAY-1S vector processor. The recursive doubling method has also been implemented on two systems, whilst the block method is currently being implemented (Barlow et al. [4]).

The implementations are fully described by Barlow et al. [2]. The results are summarised in Tables 2.1 and 2.2.

Table 2.2. ICL-DAP and CRAY-1 timings for locating a small number of eigenvalues of a matrix of size 1024.

No. of Eigen-values	ICL-DAP		CRAY-1	
	Multi-section & Bi-section	Recursive Doubling	Multi-section & Bi-section	Recursive Doubling
1	2.85 secs	1.1 secs	0.034 secs	0.145 secs
4	2.85	2.29	0.0485	0.312
16	2.85	6.8	0.0929	1.128

N.B. The coefficient matrix for the above table is the tridiagonal matrix A defined as

$$a_{ij} = \begin{cases} 2 & \text{if } |i-j| = 0 \\ -1 & \text{if } |i-j| = 1, \quad i,j = 1,2,\dots,n \\ 0 & \text{otherwise} \end{cases}$$

3. THE SOLUTION OF ORDINARY DIFFERENTIAL EQUATIONS FOR INITIAL VALUE PROBLEMS BY THE USE OF RECURRENCE RELATIONS

3.1. Initial Value Problem

The linear second order equation $y''+f(x)y'+g(x)y = k(x)$ with boundary conditions given at one end of the range only, i.e. y and y' are given or known at $x = a$. Thus, we have 2 initial conditions to start the integration procedure, proceeding to calculate y at $x_i = a+hi$, $i = 1,2,\ldots$ by a step by step process.

The derivatives are replaced by the usual central difference formulae, i.e.

$$y'' \approx (y_{r+1}-2y_r+y_{r-1})/h^2 \text{ and } y' \approx (y_{r+1}-y_r)/2h \tag{3.1}$$

to yield a 3-term recurrence equation of the form

$$(1 + \tfrac{1}{2}hf_r)y_{r+1}-(2-h^2g_r)y_r+(1 - \tfrac{1}{2}hf_r)y_{r-1}+Cy_r = h^2k_r \tag{3.2}$$

where C is a difference correction operator involving the higher order derivatives which is neglected for the initial integration of the range [a,b].

Similar methods can be applied to first order equations of the form $y'+f(x)y = k(x)$ to yield the less accurate two term recurrence relation,

$$y_{r+1}(1 + \tfrac{1}{2}hf_r)-y_r(1 - \tfrac{1}{2}hf_r)+Cy_{r+\frac{1}{2}} = \tfrac{1}{2}h(k_r+k_{r+1}). \tag{3.3}$$

For non-linear equations of the form $y'' = f(x,y)$, the recurrence relation is also generally nonlinear but some form of linearisation can be applied to obtain a linear 3 term equation of the form, i.e.,

$$y_{r+1} = b_{r+1}y_r + a_{r+1}y_{r-1} + Cy_r . \tag{3.4}$$

For the parallel evaluation of the second order recurrence relations,

$$y_{r+1} = b_{r+1}y_r + a_{r+1}y_{r-1}, \quad r = 1,2,\ldots,N \tag{3.5}$$

we can rewrite it more simply in first order form as,

$$Y_{r+1} = A_{r+1}Y_r \tag{3.6}$$

where

$$Y_{r+1} = \begin{bmatrix} y_{r+1} \\ y_r \end{bmatrix}, \quad A_r = \begin{bmatrix} b_{r+1} & a_{r+1} \\ 1 & 0 \end{bmatrix}, \quad r = 1,2,\ldots,N. \tag{3.7}$$

Then, the solution to the differential equation can be obtained as

$$Y_{r+1} = A_{r+1}Y_r = A_{r+1}A_rY_{r-1} = \ldots \prod_{i=1}^{r} A_iY_1. \tag{3.8}$$

Finally, the solution by parallel evaluation can be obtained by use of the fan-in algorithm which is depicted below as,

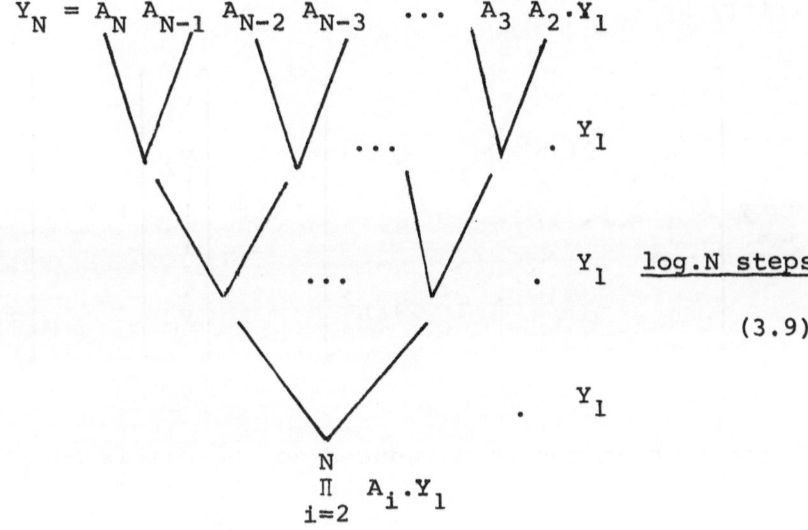

Since matrix vector multiplication is associative Y_N can be computed using the recursive doubling technique in $O(\log_2 N)$ steps.

3.2. A New Explicit Method for Two-Point Boundary Value Problems

Consider the solution of the two-point boundary value problem:

$$-\frac{d^2y(x)}{dx^2} + y(x) = f(x), \qquad 0 < x < 1, \tag{3.10}$$

with

$$y(0) = \alpha\ ,\ y(1) = \beta \tag{3.11}$$

We assume that α, β, and ρ are given constants with $\rho \geqslant 0$, and $f(x)$ is a given function such that $\frac{d^4y(x)}{dx^4}$ exists in $0 \leqslant x \leqslant 1$ and,

$$\left|\frac{d^4y(x)}{dx^4}\right| \lesssim M\ ,\quad 0 \leqslant x \leqslant 1\ . \tag{3.12}$$

Using Taylor's Theorem we can express $-\frac{d^2y}{dx^2}$ in terms of a three-point central difference approximation plus a truncation error:

$$-\frac{d^2y(x_i)}{dx^2} = [2y(x_i) - (y(x_i+h)+y(x_i-h))]/h^2 + h^2\frac{d^4y(x_i+\theta_i h)}{dx^4}/12, \tag{3.13}$$

where $|\theta_i| < 1$, $x_i \equiv ih$, $1 \leq i \leq m$, and $h \equiv 1/(m+1)$. With $y(x_i) \equiv \bar{y}_i$, the differential equation (3.10) can be written for the nodal values x_i, $1 \leq i \leq m$, in matrix form as

$$Ay \approx k \,, \tag{3.14}$$

where A is an m×m real matrix, and y and k are vectors, given explicitly by

$$A = \frac{1}{h^2}\begin{bmatrix} 2+\rho h^2 & -1 & & & 0 \\ -1 & 2+\rho h^2 & \ddots & & \\ & \ddots & \ddots & \ddots & \\ & & \ddots & \ddots & -1 \\ 0 & & & -1 & 2+\rho h^2 \end{bmatrix} ; \; y = \begin{bmatrix} y_1 \\ y_2 \\ \vdots \\ \vdots \\ y_m \end{bmatrix} ; \; k = \begin{bmatrix} f_1+\alpha/h^2 \\ f_2 \\ \vdots \\ \vdots \\ f_m+\beta/h^2 \end{bmatrix}, \tag{3.15}$$

Thus, the problem has been reduced to the matrix equation,

$$Az = k \,, \tag{3.16}$$

whose solution z is defined to be the discrete approximation to the solution $y(x)$ of (3.10)-(3.11).

We now consider a class of methods for solving the system (3.14) which is based on the "splitting" of the matrix A into the sum of three matrices,

$$A = G_1 + G_2 + \Sigma \tag{3.17}$$

where Σ is a non-negative diagonal matrix and where G_1, G_2 and Σ satisfy the following conditions:

a) $G_1+\theta\Sigma+rI$ and $G_2+\theta\Sigma+rI$ are non-singular for any $\theta \geq 0, r > 0$;
b) for any vectors c and d and for any constants $\theta \geq 0$ and $r > 0$ it is "convenient to solve the systems explicitly, i.e.,

$$x = G_1^{-1} c \qquad \text{and} \qquad y = G_2^{-1} d, \tag{3.18}$$

for x and y respectively.

We shall be concerned here with the situation where G_1 and G_2 are either small (2×2) block systems or can be made so by a suitable permutation of their rows and corresponding columns (Evans, [8]) . This procedure is "convenient" in the sense that the work required is much less than would be required to solve the original system (3.14) directly.

Now choose $h^{-1} = 5$, then for the given boundary conditions (3.11) we have

$$A = \begin{bmatrix} 2+\rho h^2 & -1 & & & \\ -1 & 2+\rho h^2 & -1 & & \bigcirc \\ & -1 & 2+\rho h^2 & -1 & \\ \bigcirc & & -1 & 2+\rho h^2 & \end{bmatrix} = G_1 + G_2 + \Sigma \tag{3.19}$$

where,

$$G_1 = \begin{bmatrix} 1 & -1 & & \bigcirc \\ -1 & 1 & & \\ & & 1 & -1 \\ \bigcirc & & -1 & 1 \end{bmatrix}, G_2 = \begin{bmatrix} 1 & & & \bigcirc \\ & 1 & -1 & \\ & -1 & 1 & \\ \bigcirc & & & 1 \end{bmatrix}, \text{ and } \Sigma = h^2\begin{bmatrix} \rho & & & \bigcirc \\ & \rho & & \\ & & \rho & \\ \bigcirc & & & \rho \end{bmatrix} \tag{3.20}$$

Let us write (3.14) in the form,

$$(G_1 + G_2 + \Sigma)y = b, \tag{3.21}$$

and let us consider two equivalent forms,

$$\left.\begin{aligned} (G_1+\theta\Sigma+rI)y &= b-(G_2+(1-\theta)\Sigma-rI)y \ , \\ (G_2+\theta\Sigma+r'I)y &= b-(G_2+(1-\theta)\Sigma-r'I)y \ . \end{aligned}\right\} \tag{3.22}$$

Analogous to the Peaceman-Rachford Method [5] one selects positive iteration parameters r and r' and determines $y^{(n+\frac{1}{2})}$ by

$$(G_1+\Sigma\theta+rI)y^{(n+\frac{1}{2})} = b-(G_2+(1-\theta)\Sigma-rI)y^{(n)} \ . \tag{3.23}$$

Then one determines $y^{(n+1)}$ by

$$(G_2+\hat{\theta}\Sigma+r'I)y^{(n+1)} = b-(G_2+(1-\hat{\theta})\Sigma-r'I)y^{(n+\frac{1}{2})} \ . \tag{3.24}$$

For simplicity we shall consider here the special case where,

$$\theta = \hat{\theta} = \frac{1}{2} \ , \quad r = r',$$

and we let

$$\bar{G}_1 = G_1 + \frac{1}{2}\Sigma \ , \quad \bar{G}_2 = G_2 + \frac{1}{2}\Sigma.$$

Evidently $\bar{G}_1$ and $\bar{G}_2$ satisfy the following conditions:

a) $\bar{G}_1 + rI$ and $\bar{G}_2 + rI$ are non-singular for any $\rho > 0$,
b) for any vectors c and d and for any $r > 0$ it is practical to solve the systems

$$(\bar{G}_1 + rI)\tilde{x} = c \ , \quad (\bar{G}_2 + rI)\tilde{y} = d$$

in explicit form since they consist of only (2×2) subsystems. Thus (3.21) becomes

$$(\bar{G}_1+\bar{G}_2)y = b. \tag{3.25}$$

and (3.22)-(3.24) become respectively,

$$(\bar{G}_1+rI)y^{(n+\frac{1}{2})} = b - (\bar{G}_2-rI)y^{(n)} \tag{3.26}$$

$$(\bar{G}_2+rI)y^{(n+1)} = b - (\bar{G}_1-rI)y^{(n+\frac{1}{2})} \tag{3.27}$$

for the Alternating Group Explicit (AGE) method.

4. NUMERICAL RESULTS

We now consider the linear problem

$$U_1' = U_2 , \tag{4.1}$$

$$U_2' = 400(U_1+\cos^2(\pi x)) + 2\pi^2\cos(2\pi x) , \tag{4.2}$$

subject to the boundary conditions

$$U_1(0) = U_1(1) = 0 . \tag{4.3}$$

The exact solution for this problem is given by,

$$\left.\begin{aligned} U_1(x) &= \frac{e^{-20}}{1+e^{-20}}.e^{20x} + \frac{1}{1+e^{-20}} \cdot e^{-20x} - \cos^2(\pi x) \\ U_2(x) &= \frac{20\ e^{-20}e^{20x}}{1+e^{-20}} - \frac{20}{1+e^{-20}}\ e^{-20x} + \pi\sin(2\pi x) \end{aligned}\right\} \tag{4.4}$$

From (4.1) and (4.2) we have

$$U_1'' = 400(U_1+\cos^2(\pi x)) + 2\pi^2 \cos(2\pi x). \tag{4.5}$$

By following the usual finite difference procedure, equation (4.5) can be approximated to obtain the linear difference equation (assuming that $u = u_1$),

$$\frac{u_{i-1}-2u_i+u_{i+1}}{h^2} = 400[u_i+\cos^2(\pi x_i)] + 2\pi^2 \cos(2\pi x_i), \quad i = 1,2,\ldots,m.$$

This equation can be simplified to the form

$$-u_{i-1}+(2+400h^2)u_i-u_{i+1} = -2h^2[200\cos^2(\pi x_i) + \pi^2\cos(2\pi x_i)] \quad i = 1,2,\ldots,m. \tag{4.6}$$

The boundary conditions are replaced by,

$$u_0 = 0 \quad \text{and} \quad u_{m+1} = 0 \tag{4.7}$$

where $h = \frac{1}{m+1}$.

The linear system (4.7) can be represented in matrix notation as

$$Au = (\bar{G}_1+\bar{G}_2)u = b \ . \tag{4.8}$$

The vector u is defined in the usual way and b is given by

$$b = (c_1,c_2,\ldots,c_{m-1},c_m)^T \ , \tag{4.9}$$

where

$$c_i = -2h^2[200\cos^2(\pi x_i) + \pi^2\cos(2\pi x_i)] \ , \ i = 1,2,\ldots,m \tag{4.10}$$

$\bar{G}_1$ and $\bar{G}_2$ are given by,

$$\bar{G}_1 = \left[\begin{array}{cc|cc|c|cc} g & -1 & & & & & \\ -1 & g & & & & & \\ \hline & & g & -1 & & & \\ & & -1 & g & & & \\ \hline & & & & \ddots & & \\ \hline & & & & & g & -1 \\ & & & & & -1 & g \end{array}\right] \text{ and } \bar{G}_2 = \left[\begin{array}{c|cc|c|cc|c} g & & & & & & \\ \hline & g & -1 & & & & \\ & -1 & g & & & & \\ \hline & & & \ddots & & & \\ \hline & & & & g & -1 & \\ & & & & -1 & g & \\ \hline & & & & & & g \end{array}\right] \tag{4.11}$$

with $g = 1+200h^2$.

Hence, by applying the A.G.E. method of (3.26-27), we can determine $u^{(n+\frac{1}{2})}$ and $u^{(n+1)}$ successively from equations (3.26)-(3.27). It is obvious that the (2×2) submatrices of $(\bar{G}_1+rI)$ and $(\bar{G}_2+rI)$ are of the form

$$\hat{G} = \begin{pmatrix} \alpha & -1 \\ -1 & \alpha \end{pmatrix}$$

where $\alpha = g+r$, and the inverse of $\hat{G}$ is given by

$$\hat{G}^{-1} = d\begin{pmatrix} \alpha & 1 \\ 1 & \alpha \end{pmatrix} \ , \quad \text{where } d = \frac{1}{\alpha^2-1} \ . \tag{4.12}$$

Hence the vector $u^{(n+1)}$ can be determined from $\underline{u}^{(n)}$ in two steps, we first determine $u^{(n+\frac{1}{2})}$ as follows,

$$\begin{bmatrix} u_1 \\ u_2 \\ u_3 \\ u_4 \\ \vdots \\ u_{m-1} \\ u_m \end{bmatrix}^{(n+\frac{1}{2})} = d \left[\begin{array}{cc|cc|c|cc} \alpha & 1 & & & & & \\ 1 & \alpha & & & & \bigcirc & \\ \hline & & \alpha & 1 & & & \\ & & 1 & \alpha & & & \\ \hline & & & & \ddots & & \\ \hline & \bigcirc & & & & \alpha & 1 \\ & & & & & 1 & \alpha \end{array}\right] \begin{bmatrix} c_1 - \beta u_1 \\ c_2 - \beta u_2 + u_3 \\ c_3 + u_2 - \beta u_3 \\ c_4 - \beta u_4 + u_5 \\ \vdots \\ c_{m-1} + u_{m-2} - \beta u_{m-1} \\ c_m - \beta u_m \end{bmatrix} \tag{4.13}$$

and by using the values of $u^{(n+\frac{1}{2})}$ we determine $u^{(n+1)}$

$$\begin{bmatrix} u_1 \\ u_2 \\ u_3 \\ \vdots \\ u_{m-2} \\ u_{m-1} \\ u_m \end{bmatrix}^{(n+1)} = \left[\begin{array}{c|cc|c|cc|c} 1/\alpha & & & & & & \\ \hline & \alpha d & d & & & \bigcirc & \\ & d & \alpha d & & & & \\ \hline & & & \ddots & & & \\ \hline & & \bigcirc & & \alpha d & d & \\ & & & & d & \alpha d & \\ \hline & & & & & & 1/\alpha \end{array}\right] \begin{bmatrix} c_1 - \beta u_1 + u_2 \\ c_2 + u_1 - \beta u_2 \\ c_3 - \beta u_3 + u_4 \\ \vdots \\ c_{m-2} + u_{m-3} - \beta u_{m-2} \\ c_{m-1} - \beta u_{m-1} + u_m \\ c_m + u_{m-1} - \beta u_m \end{bmatrix}^{(n+\frac{1}{2})} \tag{4.14}$$

where $\beta = g - r$.

The accompanying tables show the results obtained by solving the given problem by the A.G.E. method and the S.O.R. method. The convergence test used was the average test

$$||u_i^{(n+1)} - u_i^{(n)}|| / (1 + ||u_i^{(n)}||) < \varepsilon \tag{4.15}$$

with $\varepsilon = 10^{-5}$. For each method, the logarithm of the minimum number of iterations was plotted against $\log(h^{-1})$, the graph

Table 4.1.

h^{-1}	S.O.R.		A.G.E.	
	ω_b	n	r	n
13	1.075-1.077	9	1.75-1.9	5
25	1.206-1.238	16	0.58-0.98	9
37	1.325-1.374	23	0.53-0.59	12
49	1.42-1.47	30	0.25-0.465	16

which supports the A.G.E. theory is shown in Figure 4.1.

For the linear problem $\underline{u}^{(n+\frac{1}{2})}$ and $\underline{u}^{(n+1)}$ can be obtained from (4.13) and (4.14) respectively. Hence, we can show that the number of operations required to solve this problem by the A.G.E. iterative method is

$$(6m-7)\text{ multiplications} + (4m - \tfrac{11}{2})\text{ additions} + \text{R.H.S. unit} \tag{4.16}$$

per iteration.

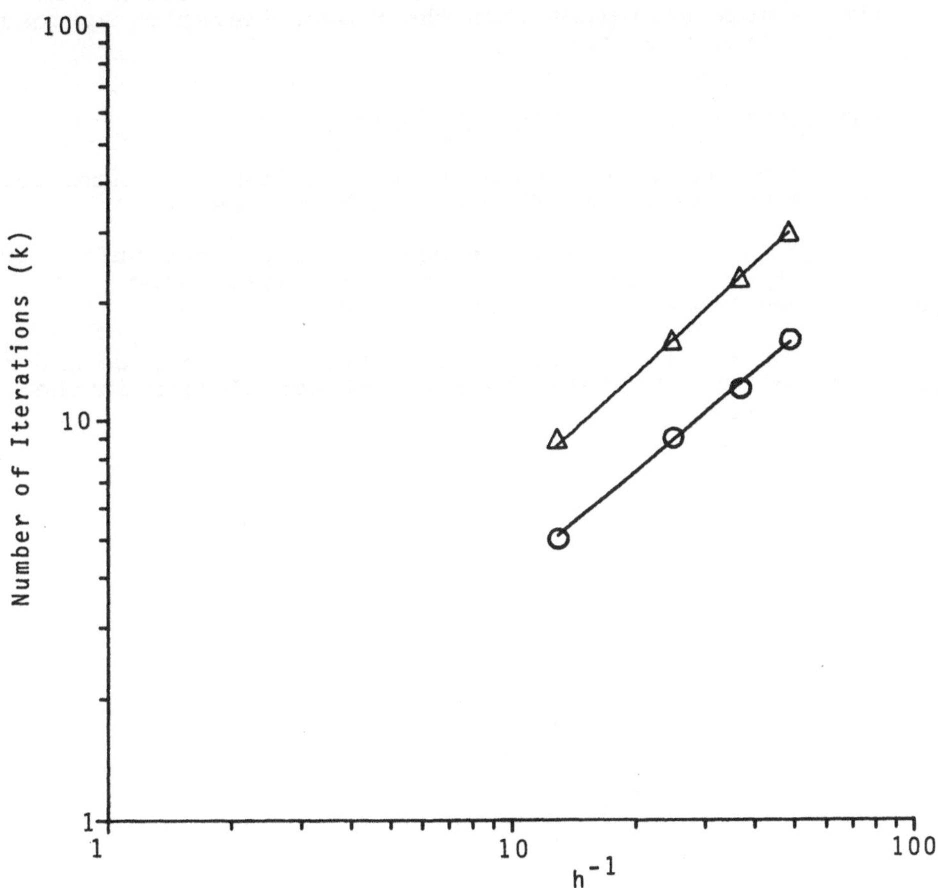

Table 4.2.

Method / h^{-1}	S.O.R.		A.G.E.	
	M	A	M	A
13	18m	36m-18	30m-35	20m-27.5
25	32m	64m-32	54m-63	36m-49.5
37	46m	92m-46	72m-84	48m-66
49	60m	120m-60	96m-112	64m-88

On the other hand, to solve this problem by the S.O.R. iterative method we require per iteration

$$2m \text{ multiplications} + (4m-2) \text{ additions} + \text{R.H.S. unit} \tag{4.17}$$

Hence, by combining the results shown in Table 4.1 with the corresponding number of operations per iteration required to solve the problem by the S.O.R. and A.G.E. methods, we can obtain the total number of arithmetic operations required.

Further it can be seen from Table 4.2 that the A.G.E. algorithm is more efficient than the S.O.R. iterative approach to solve the linear problem.

5. A FAST EXPLICIT METHOD FOR PARABOLIC EQUATIONS

Another technique of achieving parallelism in a numerical algorithm is by the use of fast explicit methods.

However these methods are among the oldest and suffer from poor stability and convergence characteristics. Also they require unacceptable computer solution times.

The newer implicit methods are better but often we are not able to exploit to the full the implicit parallelism in the solution algorithm.

Hence we must find new explicit methods with improved stability and convergence characteristics.

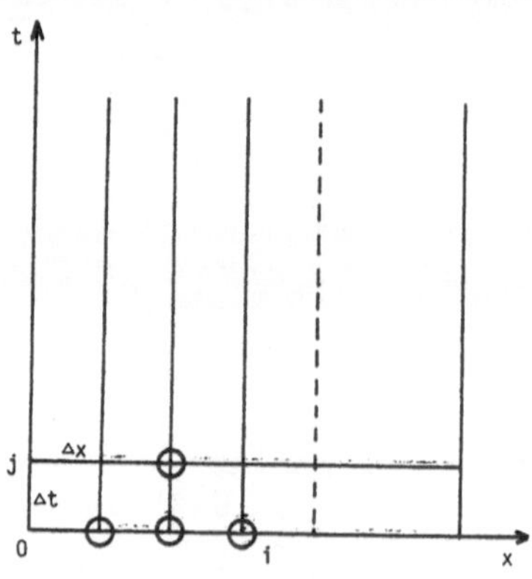

Fig. 5.1.

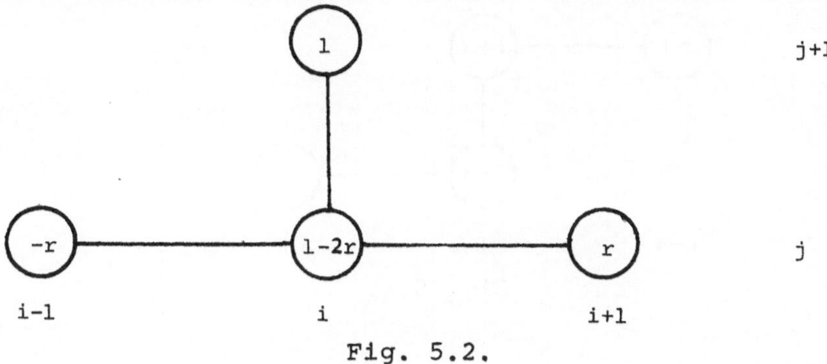

Fig. 5.2.

Consider the simple heat-conduction problem (Fig. 5.1.),

$$\frac{\partial u}{\partial t} = \frac{\partial^2 u}{\partial x^2}\ , \ 0 < x < 1,\ t > 0\ , \tag{5.1}$$

with initial conditions, $u(x,0) = f(x),\ 0 \leqslant x \leqslant 1,$

and boundary conditions, $u(0,t) = g_0(t),\ 0 < t < T\ ,$

$u(1,t) = g_1(t),\ 0 < t < T.$

The simplest explicit method uses a forward difference operator approximation to $\frac{\partial u}{\partial t}$ and a central difference operator approximation to $\frac{\partial^2 u}{\partial x^2}$. The formula,

$$u_{i,j+1} = ru_{i-1,j}+(1-2r)u_{i,j} + ru_{i+1,j} + 0(\Delta t+\Delta x^2) \tag{5.2}$$

is well-known (Fig. 5.2) but is unstable for values of $r = \frac{\partial t}{\partial x^2} > \frac{1}{2}$. Hence the algorithm is ideal for parallel application since every point on the grid can be evaluated at the same time. However the method requires long solution times due to the small time step of integration.

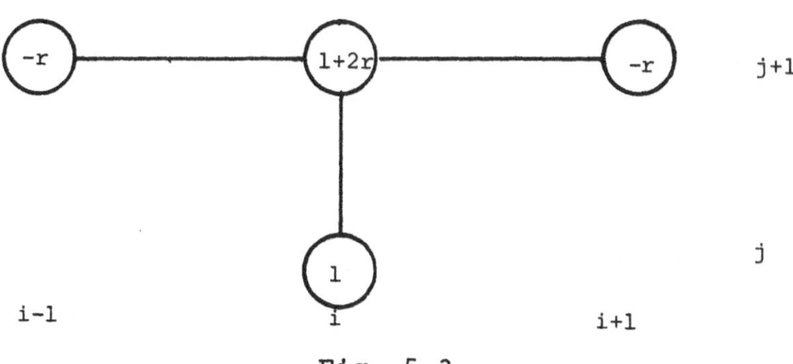

Fig. 5.3.

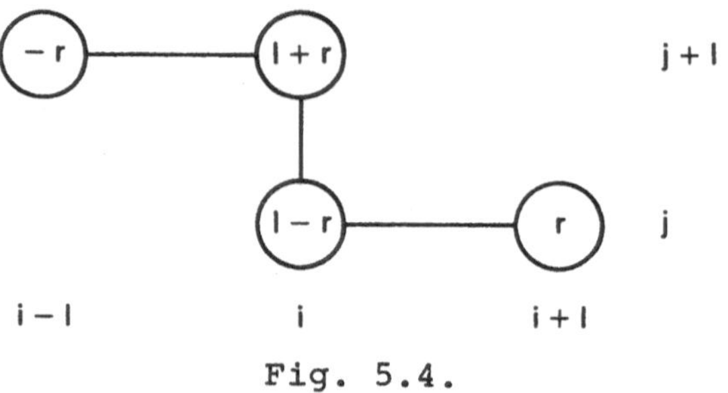

Fig. 5.4.

An implicit method uses a backward difference operator approximation to $\frac{\partial u}{\partial t}$ and a central difference operator approximation to $\frac{\partial^2 u}{\partial x^2}$. The equation,

$$-ru_{i-1,j+1}+(1+2r)u_{i,j+1}-ru_{i+1,j+1} \approx u_{i,j} \,, \tag{5.3}$$

is also well-known and is stable for all values of r (Fig. 5.3). However, the algorithm requires the solution of a system of 3 term finite difference equations at every time step in which we are not able to exploit the parallelism to the full.

In order to facilitate the solution of these implicit equations, asymmetric techniques due to Saul'yev (1964) have been used, i.e. the computational molecule Fig. 5.4 representing the equation,

$$-ru_{i-1,j+1}+(1+r)u_{i,j+1} = (1-r)u_{i,j}+ru_{i+1,j}+O(\Delta t+\Delta x^2 + \frac{\Delta t}{\Delta x}) \tag{5.4}$$

is explicit if solved from left to right and the computational molecule Fig. 5.5 representing the equation,

$$-ru_{i+1,j+1}+(1+r)u_{i,j+1} = (1-r)u_{i,j}+ru_{i-1,j}+O(\Delta t+\Delta x^2 - \frac{\Delta t}{\Delta x}) \tag{5.5}$$

is explicit if solved from right to left.

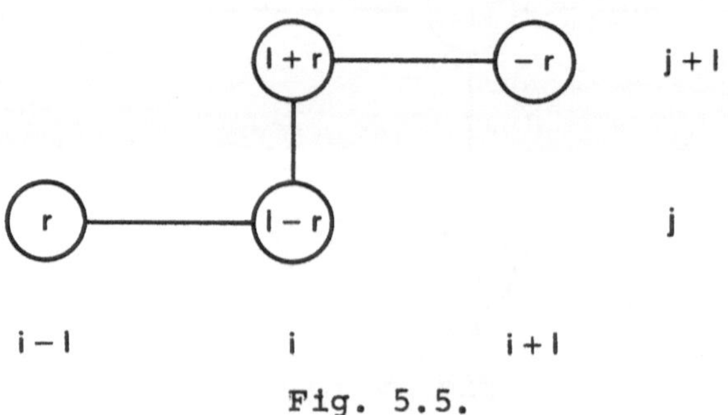

Fig. 5.5.

These two schemes are often referred to as semi-explicit formulae.

A New Group Explicit Method

If we now couple the use of the asymmetric equations (5.4) and (5.5) at 2 adjacent points (see Fig. 5.6), then they result in a (2×2) set of implicit difference equations.

For the group of two points, i.e. $\{i\Delta x,(j+\frac{1}{2})\Delta t\}$ and $\{(i+1)\Delta x,(j+\frac{1}{2})\Delta t\}$, equations (5.5) and (5.4) are used simultaneously to calculate the values of u at these points respectively. Therefore, at point $\{i\Delta x,(j+\frac{1}{2})\Delta t\}$ the solution is approximated by

$$-ru_{i+1,j+1}+(1+r)u_{i,j+1} \approx ru_{i-1,j}+(1-r)u_{i,j}, \tag{5.4a}$$

whilst at point $\{(i+1)\Delta x,(j+\frac{1}{2})\Delta t\}$, the solution is approximated by

$$-ru_{i,j+1}+(1+r)u_{i+1,j+1} = (1-r)u_{i+1,j}+r_{i+2,j}. \tag{5.5a}$$

If we now rewrite equations (5.4) and (5.5) in matrix form, we have

$$\begin{bmatrix} 1+r & -1 \\ -1 & 1+r \end{bmatrix}\begin{bmatrix} u_{i,j+1} \\ u_{i+1,j+1} \end{bmatrix} = \begin{bmatrix} 1-r & 0 \\ 0 & 1-r \end{bmatrix}\begin{bmatrix} u_{i,j} \\ u_{i+1,j} \end{bmatrix} + \begin{bmatrix} ru_{i-1,j} \\ ru_{i+2,j} \end{bmatrix} \tag{5.6}$$

in which the (2×2)matrix of coefficients can easily be inverted so that the equation can be written in explicit form as,

$$\begin{bmatrix} u_{i,j+1} \\ u_{i+1,j+1} \end{bmatrix} = \frac{1}{|A|}\begin{bmatrix} 1+r & r \\ r & 1+r \end{bmatrix}\left\{\begin{bmatrix} 1-r & 0 \\ 0 & 1-r \end{bmatrix}\begin{bmatrix} u_{i,j} \\ u_{i+1,j} \end{bmatrix} + \begin{bmatrix} ru_{i-1,j} \\ ru_{i+2,j} \end{bmatrix}\right\} \tag{5.7}$$

where $A = 1+2r$. This simplifies to

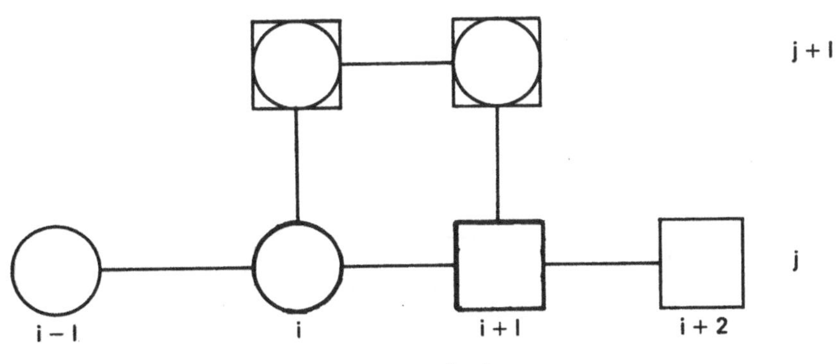

Fig. 5.6

$$\begin{bmatrix} u_{i,j+1} \\ u_{i+1,j+1} \end{bmatrix} = \frac{1}{|A|} \begin{bmatrix} r(1+r)u_{i-1,j}+(1-r^2)u_{i,j}+r(1-r)u_{i+1,j}+r^2u_{i+2,j} \\ r^2u_{i-1,j}+r(1-r)u_{i,j}+(1-r^2)u_{i+1,j}+r(1+r)u_{i+2,j} \end{bmatrix} \tag{5.8}$$

For any ungrouped (single) points near the right and left boundaries, equations (5.4) and (5.5) can be used respectively, i.e. for the right boundary,

$$u_{m-1,j+1} = \frac{1}{(1+r)} (ru_{m,j+1}+ru_{m-2,j}+(1-r)u_{m-1,j}), \tag{5.9}$$

and for the left boundary,

$$u_{1,j+1} = \frac{1}{(1+r)} (ru_{0,j+1}+ru_{2,j}+(1-r)u_{1,j}) . \tag{5.10}$$

Finally, equation (5.6) can be easily converted to explicit form resulting in the computational molecule (Fig. 5.7) representing the equation

$$u_{i,j+1} = \frac{1}{(1+2r)} [r(1+r)u_{i-1,j}+(1-r^2)u_{i,j}+r(1+r)u_{i+1,j} +r^2u_{i+2,j}] \tag{5.11}$$

and the molecule (Fig. 5.8) representing

$$u_{i+1,j+1} = \frac{1}{(1+2r)} [r^2u_{i-1,j}+r(1-r)u_{i,j}+(1-r^2)u_{i+1,j} +r(1+r)u_{i+2,j}] \tag{5.12}$$

which when used in the alternating group explicit (AGE) method results in a stable explicit algorithm which is ideal for parallel application (Evans and Abdullah, [1]).

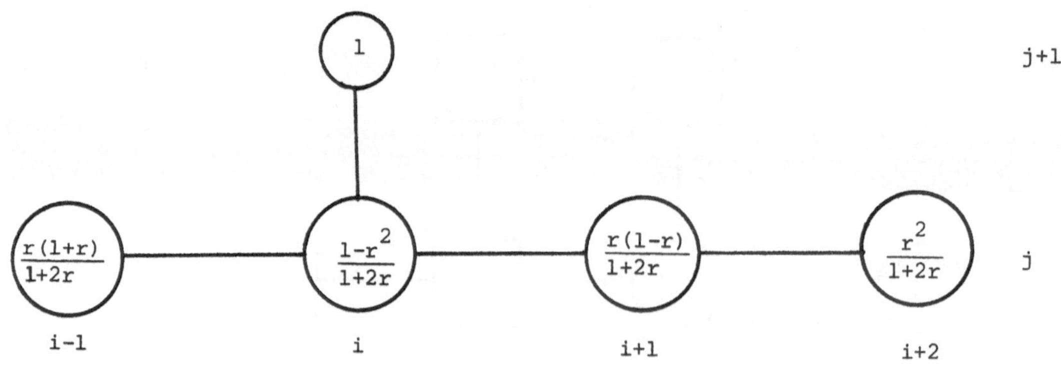

Fig. 5.7.

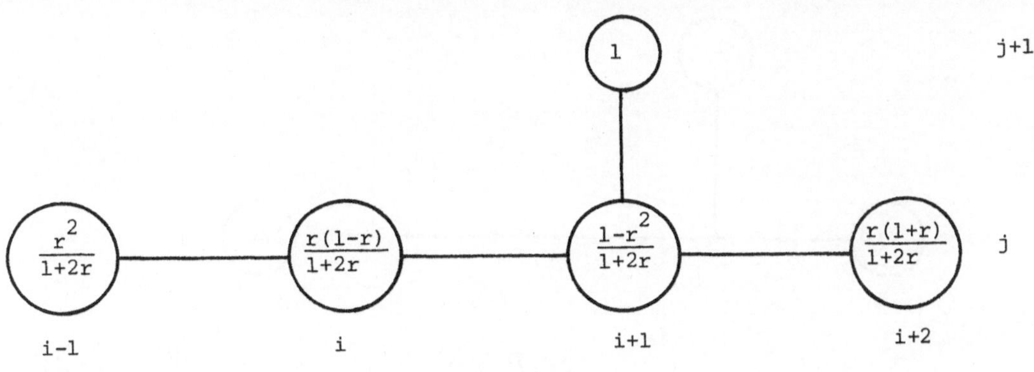

Fig. 5.8.

The given problem (5.1) was solved using the AGE algorithm on the NEPTUNE 4 processor parallel MIMD system at Loughborough University and the results obtained when compared with the standard explicit method confirm its suitability for parallel implementation.

Finally, a fast algorithmic solution can be obtained for the special value of r = 1. For this case equation (5.8) becomes,

$$\begin{bmatrix} u_{i,j+1} \\ u_{i+1,j+1} \end{bmatrix} = \frac{1}{3} \begin{bmatrix} 2u_{i-1,j} + u_{i+2,j} \\ u_{i-1,j} + 2u_{i+2,j} \end{bmatrix}$$

with the corresponding computational molecule given by Figure 5.9.

The amount of computational work required per point is just 2 additions and 1 multiplication which is less than the standard explicit method (which would be unstable at r = 1 anyway) together with the advantages of parallelisation which is derived from the explicitness.

Table 5.1.

No. of points	No. of processors	The Explicit Method		The Group Explicit Method	
		Speed-up	Efficiency	Speed-up	Efficiency
1920	0,1	1.93	0.9650	1.98	0.9900
	0,1,2	2.85	0.9500	2.95	0.9833
	0,1,2,3	3.77	0.9425	3.91	0.9775
The relative speed-up = $\frac{\text{explicit}}{\text{Group explicit}}$ =1.1619					

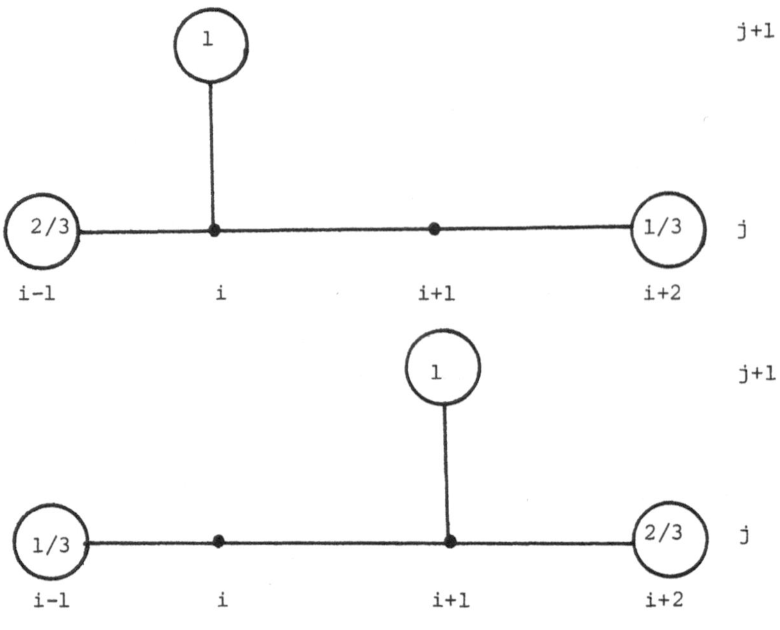

Fig. 5.9.

Finally, it can be noticed that for the Fast Group Explicit method, it is possible to obtain the solution at alternate sets of nodes (either O or .) as shown in Figure 5.9. Thus, the method (2 line Hopscotch) can advance the solution over the whole domain leaving half of the grid points uncalculated. This in effect reduces the cost of the computation by half.

6. PARALLEL ALGORITHMS FOR LINEAR SYSTEMS

Most algorithms that are found in our textbooks are based on a <u>sequential</u> way of thinking. This is undoubtedly due to the historical way of mathematical problem solving in our education and to date, the utilisation of serial computers. However, with the increasing availability of parallel computers, the 'discovery' or development of new parallel algorithms for many standard or new problems is bound to occur when <u>parallel</u> thought processes become more established.

Consider the factorisation of the matrix A in the form A = LU which forms the central theme in many linear algebra applications. The computation of the elements of L and U reduces to non-linear recurrence relations which can <u>only</u> be solved sequentially. However, Evans [9] analyses a different decomposition of the (n×n) matrix A, namely the Quadrant Interlocking factorisation (Q.I.F.), i.e.

$$A_n = W_n Z_n \, , \qquad (6.1)$$

where,

$$W_n = \left[\begin{array}{c|c|c} 1 & 0 & 0 \\ \hline w_1 & W_{n-2} & w_n \\ \hline 0 & 0 & 1 \end{array}\right] \quad \text{and} \quad Z_n = \left[\begin{array}{c|c|c} z_{1,1} & \underline{z}_1^T & z_{1,n} \\ \hline 0 & Z_{n-2} & 0 \\ \hline z_{n,1} & \underline{z}_n^T & z_{n,n} \end{array}\right], \; n \geqslant 3. \tag{6.2}$$

This decomposition leads in a natural manner to algorithms where the coefficients of the W and Z 'butterfly' matrices are obtained as (2×2) systems of linear equations which can be solved in parallel.

Consider then a factorization of the matrix A of the form,

$$A = WZ \;, \tag{6.3}$$

where,

$$W = \begin{bmatrix} 1 & \bigcirc & & 0 \\ w_{21} & 1 & 0 & w_{24} \\ w_{31} & 0 & 1 & w_{34} \\ 0 & & \bigcirc & 1 \end{bmatrix}, \text{ and } Z = \begin{bmatrix} z_{11} & z_{12} & z_{13} & z_{14} \\ \bigcirc & z_{22} & z_{23} & \bigcirc \\ & z_{32} & z_{33} & \\ z_{41} & z_{42} & z_{43} & z_{44} \end{bmatrix}. \tag{6.4}$$

In general, the matrices W and Z will have the forms,

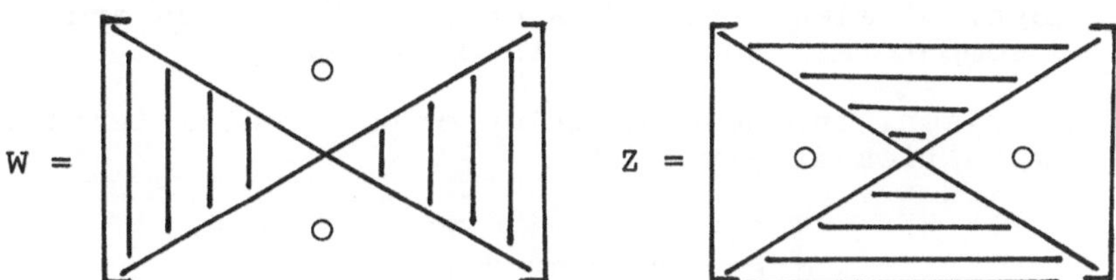

and are termed the quadrant interlocking factors (Q.I.F.) of A. It can be noticed that they have a butterfly shape (Evans [9]).

To determine the coefficients of W and Z we equate the coefficients of A and WZ in (6.4). Thus, for rows I and IV we have,

$$\begin{array}{ll} \text{I} & z_{11} = a_{11},\; z_{12} = a_{12},\; z_{13} = a_{13},\; z_{14} = a_{14}, \\ \text{IV} & z_{41} = a_{41},\; z_{42} = a_{42},\; z_{43} = a_{43},\; z_{44} = a_{44}\;. \end{array} \tag{6.5}$$

Whilst for row 2, we have the equations,

$$\begin{array}{ll} \text{II} & w_{21}z_{11}+w_{24}z_{41} = a_{21},\; w_{21}z_{12}+z_{22}+w_{24}z_{42} = a_{22}, \\ & w_{21}z_{13}+z_{23}+w_{24}z_{43} = a_{23};\; w_{21}z_{14}+w_{24}z_{44} = a_{24}\;. \end{array} \tag{6.6}$$

From the first and last equations we obtain w_{21} and w_{24} and by substitution in the 2nd and 3rd equation we obtain z_{22} and z_{23}.

Similarly for row 3, we have the equations,

$$\text{III}\quad \begin{aligned} &w_{31}z_{11}+w_{34}z_{41} = a_{31},\ w_{31}z_{12}+z_{32}+w_{34}z_{42} = a_{32}, \\ &w_{31}z_{13}+w_{34}z_{43} = a_{33},\ w_{31}z_{14}+w_{34}z_{44} = a_{34}\ . \end{aligned} \tag{6.7}$$

As before we obtain from the first and last equations the values of w_{31} and w_{34} and by substituting in the 2nd and 3rd equations, we obtain z_{32} and z_{33}.

Thus, we can see that the first and last rows of Z are given immediately. Then, (2×2) sets of linear equations are solved to obtain

$w_{i,1}$ and $w_{i,4}$ for $i = 2,3$.

Thus, the calculation proceeds as follows

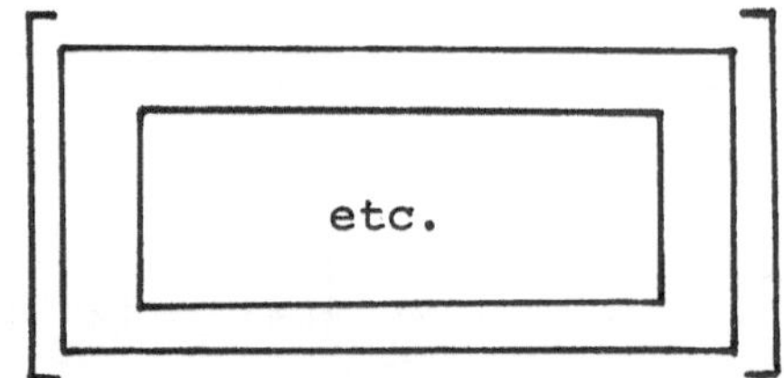

where the outermost peripheral elements of the matrices W and Z are obtained. Then, the calculation proceeds to the innermost next layer of elements. Thus only $(\frac{n-1}{2})$ stages are required to compute all the elements of W and Z.

In comparison, the determination of the coefficients in the LU decomposition is given as

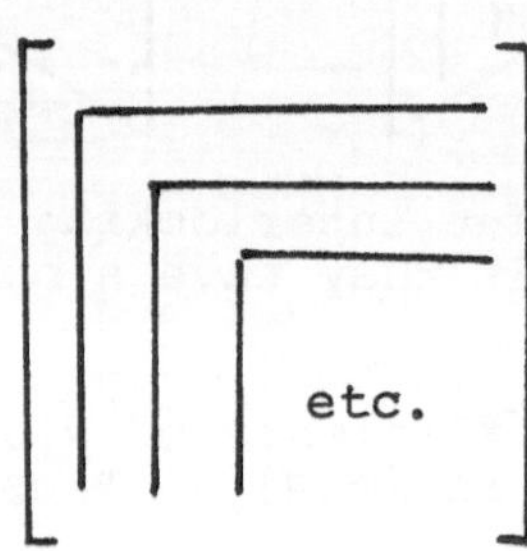

Solution of the Linear Systems

Using the relationship

$A = WZ$,

then the linear system

$Ax = b$,

can now be reformulated as the solution of 2 related linear systems,

$$Wy = b \ , \quad \text{and} \quad Zx = y \ . \tag{6.8}$$

To solve $Wy = b$ we proceed as follows:

$$\begin{bmatrix} 1 & & \bigcirc & 0 \\ w_{21} & 1 & 0 & w_{24} \\ w_{31} & 0 & 1 & w_{34} \\ 0 & & \bigcirc & 1 \end{bmatrix} \begin{bmatrix} y_1 \\ y_2 \\ y_3 \\ y_4 \end{bmatrix} = \begin{bmatrix} b_1 \\ b_2 \\ b_3 \\ b_4 \end{bmatrix}$$

We see immediately that,

$$\begin{gathered} y_1 = b_1 \quad \text{and} \quad y_4 = b_4, \\ w_{21}y_1+y_2+w_{24}y_4 = b_2 \quad \text{and} \quad w_{31}y_1+y_3+w_{34}y_4 = b_3, \end{gathered} \tag{6.9}$$

or

$$y_2 = \tilde{b}_2 = (b_2-w_{21}y_1-w_{24}y_4),$$

and

$$y_3 = \tilde{b}_3 = (b_3-w_{31}y_1-w_{34}y_4) \ .$$

The solutions for y are obtained in pairs working from the top and bottom components of the vector.

Once the vector y has been determined then to solve the system $Zx = y$ we proceed as follows,

$$\begin{bmatrix} z_{11} & z_{12} & z_{13} & z_{14} \\ & z_{22} & z_{23} & \\ & z_{32} & z_{33} & \\ z_{41} & z_{42} & z_{43} & z_{44} \end{bmatrix} \begin{bmatrix} x_1 \\ x_2 \\ x_3 \\ x_4 \end{bmatrix} = \begin{bmatrix} y_1 \\ y_2 \\ y_3 \\ y_4 \end{bmatrix}$$

Starting at the centre we solve the (2×2) linear system,

$$\begin{aligned} z_{22}x_2 + z_{23}x_3 &= y_2 \ , \\ z_{32}x_2 + z_{33}x_3 &= y_3 \ , \end{aligned} \tag{6.10}$$

to evaluate x_2 and x_3.

Then we proceed outwards and solve the (2×2) linear system,

$$\begin{aligned} z_{11}x_1+z_{14}x_4 &= \tilde{y}_1 = (y_1-z_{12}x_2-z_{13}x_3) \\ z_{41}x_1+z_{44}x_4 &= \tilde{y}_4 = (y_4-z_{42}x_2-z_{43}x_3) \end{aligned} \tag{6.11}$$

to evaluate x_1 and x_4.
Thus, the solution x can be obtained in O(n) stages on a Parallel Computer with $O(n^2)$ processors.

Thus, for general (n×n) matrices, a factorisation of the form

$$A = WZ \;,$$

is possible where W and Z have the matrix forms,

$$W = \begin{bmatrix} 1 & & & & & & 0 \\ w_{2,1} & 1 & & \circ & & & w_{2,n} \\ \vdots & w_{3,2} & \ddots & & & \vdots & \vdots \\ \vdots & \vdots & & 1 & & \vdots & \vdots \\ \vdots & w_{n-2,2} & & & \ddots & & \vdots \\ w_{n-1,1} & & & \circ & & & w_{n-1,n} \\ 0 & & & & & & 1 \end{bmatrix}, \quad Z = \begin{bmatrix} z_{1,1} & z_{1,2} & \cdots & \cdots & z_{1,n} \\ & z_{2,2} & \cdots & z_{2,n-1} & \\ \circ & & z_{i,i} & & \circ \\ & \ddots & \cdots & \ddots & \\ z_{n,1} & \cdots & \cdots & \cdots & z_{n,n} \end{bmatrix}$$

where the elements of W and Z are given by

$$W_{i,j} = \begin{cases} 1, & i=j, \\ 0, & i=1(1)\frac{n+1}{2},\; j=i+1(1)n-1,1, \\ 0, & i=\frac{n+2}{2}(1)n,\; j=n-i+1(1)i-1, \\ w_{i,j} & \text{otherwise} \end{cases} \qquad Z_{i,j} = \begin{cases} z_{i,j}, & i=1(1)\frac{n+1}{2} \\ z_{i,j}, & i=\frac{n+2}{2}(1)n, \\ & j=n-i+1(1)i \\ 0, & \text{otherwise} \end{cases}$$

<u>Computation of the Matrices W and Z</u>

By comparing terms of A and WZ we have:

1. the elements of the first and last row of Z are given immediately:

$$z_{1,i} = a_{1,i} \quad \text{and} \quad z_{n,i} = a_{n,i} \quad \text{for all } i = 1(1)n$$

2. then the sets of (2×2) linear systems given by,

$$\begin{aligned} z_{1,1}w_{i,1} + z_{n,1}w_{i,n} &= a_{i,1} \;, \\ z_{1,n}w_{i,1} + z_{n,n}w_{i,n} &= a_{i,n} \;, \end{aligned} \qquad (6.12)$$

are solved to obtain the values of $w_{i,1}$ and $w_{i,n}$ for $i = 2(1)n-1$. This then completes the <u>first stage</u> and the calculation of the outermost elements of the matrices W and Z.

At least $(\frac{n-1}{2})$ such stages are required to compute all the elements of the matrices W and Z.

Solution of the Linear System

By using the relationship

$$A = WZ ,$$

the linear system $A\underline{x} = \underline{b}$ can be reformulated as the solution of the 2 related linear systems

$$Zx = y \quad \text{and} \quad Wy = b = s$$

These are linear systems of the form,

$$\begin{bmatrix} 1 & & & & 0 \\ w_{2,1} & 1 & \circ & & w_{2,n} \\ \vdots & w_{3,2} & \ddots & \iddots & \vdots \\ \vdots & \vdots & 1 & & \vdots \\ \vdots & \iddots & & \ddots & \vdots \\ w_{n-1,1} & & & & w_{n-1,n} \\ 0 & & \circ & & 1 \end{bmatrix} \begin{bmatrix} y_1 \\ y_2 \\ \vdots \\ \vdots \\ \vdots \\ y_{n-1} \\ y_n \end{bmatrix} = \begin{bmatrix} s_1 \\ s_2 \\ \vdots \\ \vdots \\ \vdots \\ s_{n-1} \\ s_n \end{bmatrix}$$

We see that y_1 and y_n are calculated first then y_2, y_{n-1} and so on in pairs working from the top and rear of the vector y and s = b.

In general, at the ith step we have,

$$y_i = s_i, \; y_{n-i+1} = s_{n-1+1} \, , \tag{6.13}$$

and we reset the s_j in the following manner,

$$s_j = s_j - w_{j,i} y_i - w_{j,n-i+1} y_{n-i+1}, \; j = i+1(1)n-i. \tag{6.14}$$

Similarly, the system Zx = y can be treated in an analogous manner.
For parallel computers with $0(n^2)$ processors - this is an O(n) method.

Finally, it can be shown that by suitably chosen permutation matrices the method is identical to a (2×2) block Gaussian Elimination technique.

7. THE PARALLEL SOLUTION OF BANDED LINEAR SYSTEMS

The solution of Banded and Tridiagonal linear systems is a commonly occurring problem in Computational Mathematics. Here we compare the existing algorithm with a new parallel strategy for n = 6 [9].

Sequential algorithms

To solve Tx = d, where T is a $(n \times n)$ tridiagonal matrix (diagonally dominant), x and d are (n×1) vectors which comprise the unknown solution and known righthand side respectively.

Initial form $\qquad\qquad Tx = d,$
i.e.,

$$\begin{bmatrix} b_1 & c_1 & & & & \\ a_1 & b_2 & c_2 & & & \bigcirc \\ & a_3 & b_3 & c_3 & & \\ \searrow & & a_4 & b_4 & c_4 & \\ & & & a_5 & b_5 & c_5 \\ \bigcirc & & & & a_6 & b_6 \end{bmatrix} \begin{bmatrix} x_1 \\ x_2 \\ x_3 \\ x_4 \\ x_5 \\ x_6 \end{bmatrix} = \begin{bmatrix} d_1 \\ d_2 \\ d_3 \\ d_4 \\ d_5 \\ d_6 \end{bmatrix}$$

Stage I: We proceed with forward elimination processes to achieve the upper triangular form Ux = h. This results in the computation:

$$\begin{aligned} &i = 1, && g_1 = c_1/b_1, && h_1 = d_1/b_1 \\ &i = 2(1)5, && g_i = c_i/(b_i-a_ig_{i-1}), && h_i = (d_i-a_ih_{i-1})/(b_i-a_ig_{i-1}) \\ &i = 6, && g_6 = 0, && h_6 = (d_6-a_6h_5)/(b_6-a_6g_5) \end{aligned} \qquad (7.1)$$

The final upper triangular form is achieved thus,

$$\begin{bmatrix} 1 & g_1 & & & & \\ & 1 & g_2 & & \bigcirc & \\ & & 1 & g_3 & \nwarrow & \\ & & & 1 & g_4 & \\ \bigcirc & & & & 1 & g_5 \\ & & & & & 1 \end{bmatrix} \begin{bmatrix} x_1 \\ x_2 \\ x_3 \\ x_4 \\ x_5 \\ x_6 \end{bmatrix} = \begin{bmatrix} h_1 \\ h_2 \\ h_3 \\ h_4 \\ h_5 \\ h_6 \end{bmatrix}$$

State II: We now proceed with a back substitution procedure to yield the solution,

$$\begin{aligned} &i = 6 && x_6 = h_6, \\ &i = 5(-1)1, && x_i = h_ig_ix_{i+1} \;. \end{aligned} \qquad (7.2)$$

This serial algorithm involves recursive sequences (7.1) and (7.2) which are not suitable for parallel computation.

Parallel Algorithm

We now partition the matrix into 2 subproblems in which the elimination processes are carried out simultaneously from the top and bottom of the matrix. Thus we have,

$$Tx = d,$$

i.e.,

$$\begin{bmatrix} b_1 & c_1 & & & & \\ a_2 & b_2 & c_2 & & \bigcirc & \\ & a_3 & b_3 & c_3 & & \\ & & a_4 & b_4 & c_4 & \\ \bigcirc & & & a_5 & b_5 & c_5 \\ & & & & a_6 & b_6 \end{bmatrix} \begin{bmatrix} x_1 \\ x_2 \\ x_3 \\ x_4 \\ x_5 \\ x_6 \end{bmatrix} = \begin{bmatrix} d_1 \\ d_2 \\ d_3 \\ d_4 \\ d_5 \\ d_6 \end{bmatrix}$$

resulting in the intermediate form, in which there is a small subssystem (i.e. 2×2) at the centre of the matrix which needs to be solved independently:

Stage I: We proceed with parallel forward and backward elimination processes to achieve the form Sx = h. (S here represents a "mnemonic' which represents the final form of the matrix. This results in the computation:

$i = 1,\quad g_1 = c_1/b_1,\ h_1 = d_1/b_1$; $j = 6\quad f_n = a_6/b_6,\ h_6 = d_6/b_6$

$i = 2(1)3,\quad g_i = c_i/(b_i - a_i g_{i-1}),$; $j=5(-1)4,\ f_j = a_j/b_j - c_j f_{j+1}),$

$h_i = (d_i - a_i h_{i-1})/(b_i - a_i g_{i-1});$ $\quad h_j = (d_j - c_j h_{j+1})/(b_j - c_j f_{j+1})$

$$\begin{bmatrix} 1 & g_1 & & & & \\ & 1 & g_2 & & \bigcirc & \\ \underline{\text{solve}} \rightarrow & & 1 & g_3 & & \\ & & f_4 & 1 & & \\ \bigcirc & & & f_5 & 1 & \\ & & & & f_6 & 1 \end{bmatrix} \begin{bmatrix} x_1 \\ x_2 \\ x_3 \\ x_4 \\ x_5 \\ x_6 \end{bmatrix} = \begin{bmatrix} h_1 \\ h_2 \\ h_3 \\ h_4 \\ h_5 \\ h_6 \end{bmatrix}$$

Stage II: Solution of central subsystem

$$\begin{bmatrix} 1 & g_3 \\ f_4 & 1 \end{bmatrix} \begin{bmatrix} x_3 \\ x_4 \end{bmatrix} = \begin{bmatrix} h_3 \\ h_4 \end{bmatrix}$$

The solution of this subsystem can be obtained by a simple

elimination process, i.e.,

$$x_3 = (h_3-g_3h_4)/(1-g_3f_4),\quad x_4 = (h_4-f_4h_3)/(1-g_3f_4)$$
$$= \bar{h}_3 \qquad\qquad\qquad\qquad = \bar{h}_4$$

The final form is then achieved as

$$Sx = h$$

$$\begin{bmatrix} 1 & g_1 & & & & \\ & 1 & g_2 & & & \\ & & 1 & & & \\ & & & 1 & & \\ & & & f_5 & 1 & \\ & & & & f_6 & 1 \end{bmatrix} \begin{bmatrix} x_1 \\ x_2 \\ x_3 \\ x_4 \\ x_5 \\ x_6 \end{bmatrix} = \begin{bmatrix} h_1 \\ h_2 \\ h_3 \\ h_4 \\ h_5 \\ h_6 \end{bmatrix}$$

Stage III: We now proceed with parallel forward and backward substituion procedures in each subproblem to yield the solution, i.e.,

$$i = 3, \quad x_3 = \bar{h}_3 \quad , j = 4, \quad x_4 = \bar{h}_4$$
$$i = 2(-1)1, \quad x_i = h_i g_i x_{i+1} \quad , j = 5(1)6, \quad x_j = h_j - f_j x_{j-1},$$

Finally, the situation for systems of equations of odd order, i.e. n = 7, can be treated in a similar manner except that the subsystem in Stage II is now not centrally situated and the two elimination half processes are not exactly equal resulting in a slighly unbalanced computation.

8. REFERENCES

1. R.H. Barlow and D.J. Evans, 'A parallel organization of the bisection algorithm', Comp.J., 22:267-269 (1978).
2. R.H. Barlow, D.J. Evans and J. Shanehchi, 'Parallel multisection applied to the eigenvalue problem', Comp.J., 26:6-9 (1983).
3. J.J. Lambiotte, 'The solution of linear systems of equations on a vector computer', Ph.D. Thesis, Univ. of Virginia (1975).
4. R.H. Barlow, D.J. Evans and J. Shanehchi, 'Comparison of parallelism between a block and multisection method for eigenvalues', Comp. Stud. Rep. 174, L.U.T.
5. D.W. Peaceman and H.H. Rachford, Jr., 'The numerical solution of parabolic and elliptic equations', J. Soc. Ind. App. Math., 3:28-41 (1955).
6. V.K. Saul'yev, 'Integration of Equations of Parabolic Type by the Method of Nets', MacMillan, New York (1964).

7. D.J. Evans and A.R.B. Abdullah, 'A new explicit method for the diffusion equation', in: 'Numerical Methods in Thermal Problems III', R.W. Lewis et al., eds., Pineridge Press (1983), p. 330-347.
8. D.J. Evans, 'New parallel algorithms for partial differential equations', in: 'Parallel Computing 83', M. Feilmeier, G.R. Joubert and U. Schendel, eds., Elsevier Publ. (1984), p. 3-56.
9. D.J. Evans, 'New parallel algorithms in linear algebra', in: 'Calcul Vectoriel et Parallèle', A. Bossavit, ed., Bulletin de la Direction des Studies et Recherches, Electricité de France (1983), p. 61-69.

DEVELOPMENTS IN SUPERCOMPUTER LANGUAGES

R.H. Perrott

Department of Computer Science, The Queen's University, Belfast BT7 1NN, Northern Ireland

ABSTRACT

Concurrent languages, such as Ada and Pascal Plus, have been designed and implemented for configurations which consist of a number of independent and concurrently operating processors. The development of languages for programming array and vector processors has proceeded independently and produced variants of Fortran for representing this type of parallel processing. However, the latest hardware configurations contain both types of parallelism, for example, the Cray X-MP contains several vector processors which are capable of acting independently and in parallel. It is appropriate with the introduction of these new configurations to consider the design of a language capable of handling both types of parallelism. Such a proposal is considered in this article.

1. INTRODUCTION

The developments in hardware technology which have enabled the production of supercomputers are well documented and well utilized in existing systems. However, in the case of the software for these supercomputers the developments have not been so radical or widely accepted among the user community. One of the reasons for this reluctance to accept new software techniques is that these machines have been and are widely used by the engineering and scientific communities. Over the last thirty years, these communities have been well served by a useful and enduring tool in the Fortran language.

This restriction to a Fortran base has, in turn, affected the languages which are proposed or implemented for these new architectures. Originally these machines consisted of an array or vector processor, but now configurations are being produced which consist of array or vector processors as part of a multiprocessing system. As a result languages are required which capitalize on the array or vector processing facilities as well as the other components of a multiprocessing system.

At The Queen's University we have been carrying out research into the design and implementation of parallel languages which are suitable for both array and vector processor configurations and multiprocessor configurations. The array and vector processor work originated in a project carried out at the NASA Ames Research Center in California, to design a parallel language for the successor to the Illiac IV. This language is known as Actus (Perrott, 1976).

Since that time several language experiments have been carried out based on the philosophy expounded in the Actus language. This has led to a gradual evolution and better understanding of language features suitable for array and vector processing. Other colleagues have been investigating multiprocessor programming languages and have produced the language Pascal Plus (Welsh and Bustard, 1979) based on the monitor and conditions for process synchronization. As a result of these experiences a recent project has considered the synthesis of these two languages to produce a language for programming a configuration which can consist of array or vector processors and other independent processors all executing in parallel (Orr, 1986).

2. MULTIPROCESSING/DISTRIBUTED PROGRAMMING

The term process or more recently task is used to describe a sequence of program instructions that can be performed in parallel with other program instructions. A program can therefore be represented as a number of processes which can be executing concurrently. The point at which a processor is withdrawn from one process and given to another is dependent on the progress of the processes and the algorithm used to assign the available processor(s). The net effect is that processes are capable of interacting in a time-dependent manner.

Thus in a concurrent programming environment a programmer requires not only program and data structures similar to those required in a sequential programming environment but also tools to control the interaction of the processes - processors which are proceeding at fixed but unknown rates.

The situations in which processes interact can be divided into two categories. The first situation occurs whenever processes wish to update a shared variable or a resource at the same time. For example, when several processes wish to use the same resource, only one process must succeed in gaining access to the resource at any time. Once a process has obtained the resource it must be able to use the resource without interference from the other competing processes. This is described as mutual exclusion.

The second situation occurs when processes are co-operating, they must be correctly synchronized with respect to each other's activities. For example, when one process requires a result not yet produced by another process. The first process must be able to wait on the second process and the second process must take the responsibility of resuming the first process when it arrives with the result. The processes are communicating or scheduling one another and are aware of each other's existence and purpose. This is described as process synchronization.

Two main techniques which have been proposed for the solution of these problems. The first technique is based on a monitor construct plus condition variables. In this solution the processes deposit shared information in a data structure and synchronize each other by means of queues using special operators. The second technique is based on processes passing information directly when they wish to communicate. There is no shared data structure and such a technique is referred to as a message passing technique.

Languages have been developed using both techniques. For example, Concurrent Pascal by Brinch Hansen (1974), Modula by Niklaus Wirth (1977) and Pascal Plus (Welsh and Bustard, 1979), all these languages are based on the monitor plus condition variable approach. To enable the user to control parallelism these languages provide a process which consists of a private data structure and a sequential program that can operate on the data. One process cannot operate on the private data of another process. A monitor defines a shared data structure and all the operations that can be performed on it. These operations are defined by the procedures of the monitor. In addition, a monitor defines an initialization operation that is executed when its data structure is created.

In general this process can access the shared data of a monitor by calling one of the monitor's procedures. If there is more than one call then only one of the calling processes is allowed to succeed in entering the monitor at any time; to guarantee that the data of the monitor are accessed exclusively. Only when a process exits the monitor is it safe for one of the calling processes which was delayed to enter the monitor.

It is also possible for a process to enter the monitor and discover that the information it requires has not yet arrived. In such a situation, it can join a queue associated with that condition and thereby release its exclusive access over the monitor. Eventually, another process may enter the monitor and enable a delayed process to continue. The queues within a monitor are usually identified by condition variables and a process can append itself to a single condition variable queue by executing a wait operation. Another process executing a signal operation on a condition variable queue will cause a process delayed on that queue (if there is one) to be resumed. These are the main additions to sequential Pascal to enable parallel programming to be specified. They are illustrated in the following code which shares a single resource among a number of processes:

```
monitor RESOURCE;
var FREE:BOOLEAN;
instance
   BUSY:CONDITION
procedure*ACQUIRE;
   begin
      if not FREE then BUSY.WAIT;
      FREE := FALSE
   end (* acquire *);
```

```
procedure *RELEASE;
   begin
      FREE := TRUE;
      BUSY.SIGNAL
   end (* release *);
begin
   FREE := TRUE;
end (* resource *);

process PRODUCER;
   (* local data *)
   begin
      (* statements *)
      RESOURCE.ACQUIRE;
      (* use resource *)
      RESOURCE;RELEASE;
      (* statements *)
   end;
instance
   BEES:array[1..N] of PRODUCER;
```

The last declaration causes N producer processes to be created with a life style defined by the above code. The N processes operate in parallel and independently making calls to the monitor whenever they wish to acquire and release a resource.

3. LANGUAGE APPROACHES

There have been three main approaches used for supercomputer languages (Perrott, 1987), namely,

Detection of Parallelism Languages

In this language group the programmer uses a sequential programming language, usually Fortran, for the application and it is the responsibility of the compiler to determine which parts of the application can be executed in parallel on the existing hardware. The major motivation for such an approach is to utilize existing sequential programs and consequently save on development costs. However, to gain any substantial benefit the user must restructure the code, in the case of existing programs, or when designing a new program take into account the direction mechanism of the compiler. This has been a widely used technique for supercomputers and has been particularly successful in the case of the Cray-1 and the more recent Japanese supercomputers from Fujitsu, NEC and Hitachi.

Expression of Machine Parallelism Languages

In this group the languages provide either a syntax which directly reflects the architecture of the machine or they demand that the programmer explicitly encodes hardware instructions in separate subroutine calls. This effectively turns these languages into high level assembly languages. Their major advantage is that the implementation problems, which are a major challenge, are considerably simplified. Examples of such languages have been mainly restricted to array processors, such as the Illiac IV and the ICL Distributed Array Processor. In these languages the size of the grid or array of processors is reflected in the syntax and the user must squeeze the problem into that size to use the hardware efficiently.

Expression of Problem Parallelism Languages

A third approach has been investigated which aims to exploit the parallelism in a problem. In this case the data and program structures of the language enable a programmer to express directly the parallel nature of a problem without reference to the hardware or the detection mechanism of the compiler. The major advantage of this approach is that the language is suitable for implementation on both array and vector processor configurations. This, in turn, should enable some measure of program portability to be achieved. An example of this language type is Actus; this is also the approach being followed in the proposed new Fortran standard.

4. LANGUAGE FEATURES

The approach proposed for configurations which consist of vector processors and other independent processors is to use an existing language such as Pascal Plus (in our case) and to have embedded within the process concept the possibility of processes which are intended to operate on a vector processor. Such processes are then co-ordinated using the language's synchronization primitives, namely, monitors and conditions. The following sections therefore describe the array and vector processing features which are intended to be used to define one of these processes.

Data Declarations

The array is used to declare and to indicate how much parallelism can be applied to that structure in the course of program execution by array and vector processors. This is referred to as the maximum extent of parallelism. Each array when it is declared has its own extent of parallelism associated with it which can be subsequently manipulated by the program statements. Each time a parallel structure is referred to in an expression or statement its extent of parallelism must be less than or equal to that associated with it at declaration time. Examples are as follows:

```
var
  vector_a: array [1:n] of integer;(* vector *)
  parallel_a: array [1:n,1:n] of real;(* parallel matrix *)
  sequential_a: array [1..n, 1..n] of real;(* scalar
                                               matrix *)
  mixed_a: array [1..n,1..n] of integer;
```

The general mechanism for selecting elements of an array is called an index set. It may be regarded as a boolean mask which is superimposed upon an array to select the required elements for processing. For example,

```
one_dim_is: indexset of 1:100;
two_dim_is: indexset of 1:100, 1:100;
```

An N-dimensional array can be processed in parallel using an N-dimensional index set to select those elements to be processed in parallel.

Statements

All operations on parallel structures are performed on an element by element basis and all operations on a parallel structure take place within a forall statement. The forall serves both to initialize an index set and to delineate its scope. It has the basic form:

```
forall index_set in initial value d o
statement part
```

More specifically

```
forall twodim is in [1:50, 1:50] d o
parallel_a[two_dim_is] := 0,0 ;
```

will initialize the upper left hand submatrix of the array parallel_a.

In addition to this basic form there are two further derivatives of the forall statement which are provided to handle parallel selection and repetition. In the first form the forall statement can be used to select items which satisfy some condition. The condition is inserted in the forall statement by means of a boolean expression. This, in turn, selects the required indices from the index set initialized in the forall statement. As a consequence the boolean expression must return a result for every member of the index set involved. For example, given the declarations

```
var
   a: array [1:200] of real;
   is: indexset of 1:100;
```

then the statement

```
forall is in [1:100] d o
   where a[is] > epsilon d o
      statement-part
```

means that the index set is initialized to the values 1 to 100 and then a test for those elements of the array a greater than epsilon is applied. Hence in the statement part the index set consists of those indices formed using both criteria.

The second form of the forall statement involves repetition. This statement has the effect of repeatedly executing the statement part until all the selected items obey the specified condition. For example

```
forall is in [1:100] d o
   aslongas a[is] > epsilon d o
      statement-part
```

means that the statement part will be repeatedly executed while any elements of a are greater than epsilon.

In order to construct more complex processing patterns two special binding operators called perm and join are provided. The perm operator has the effect of creating an index set from all the possible permutations of its two index operands. For example

```
[1:3 perm 1:2]
```

sets up the two-dimensional index set

```
[(1,1), (1,2), (2,1), (2,2), (3,1), (3,2)]
```

and when used in a two-dimensional parallel array references the elements indicated.

The join operator has the effect of pairing elements of the index operands. For example

```
[1:4 join 4:(-1)1]
```

establishes the index set (1,4), (2,3), (3,2), (4,1). The -1 in parentheses indicates a decrement value.

Data Movement

Since array and vector processors have been devised to handle large amounts of data using synchronized operations they do not suffer from the synchronization problems of multiprocessor machines. However the problems which array and vector processors do have involve the movement of data between different parallel processing streams or within the same parallel processing stream. There are two types of operations required. The first is specified by means of a shift operator which has the effect of shifting the values of the indices in the index set along their base sequence by the corresponding expression amount. For example, given the declarations

```
var
   aa: array [1:3,1:3] of integer;
   is: indexset of 1:2, 1:2;
```

the statement

```
aa[is] := aa[is shift (1,1)]
```

has the effect of making the parallel assignments

```
aa[1,1] := aa[2,2] ; aa[1,2] ; aa[2,3] ;
aa[2,1] := aa[3,2] ; aa[2,2] ; aa[3,3]
```

The second data movement operation is performed by means of a rotate operator and moves the data within the index set specification in a circular manner. The sign of the expression indicates whether the movement is forward or backward. For example, given the declarations

```
var
   aa: array [1:3,1:3] of integer;
   is: indexset of 1:2, 1:2;
```

the statement

```
aa[is] := aa[is rotate (1,1)]
```

performs the parallel assignments

```
aa[1,1] := aa[2,2] ; aa[1,2] ; aa[2,1] ;
aa[2,1] := aa[1,2] ; aa[2,2] ; aa[1,1]
```

These are just some of the language features which have been introduced to handle the array or vector processing parts of a parallel configuration. Other details can be found in Orr (1986) and Perrott and Orr (1987).

5. CONCLUDING REMARKS

The above language features are specifically associated with processing array and vector processors and enable the user to specify a solution to a problem in a manner which is independent of the machine architecture or the detection mechanism of a compiler. These parallel data structures and program structures enable a user to exploit the parallelism which is inherent in an application.

If these constructs are included in a language which provides asynchronous parallel features such as Ada and Pascal Plus then it is possible to construct a language capable of handling the two types of parallelism which are now appearing in the latest hardware configurations.

REFERENCES

Brinch Hansen, P., 1975, The programming language concurrent Pascal, IEEE Trans. on Soft. Eng., 1:199-207.

Orr, R., 1986, "Language Extensions for Array and Multiprocessor Configurations", Ph.D. Thesis, Queen's University, Belfast.

Perrott, R.H., 1979, A language for vector and array processors, ACM TOPLAS, 1:177-195.

Perrott, R.H., 1987, "Parallel Programming", Addison Wesley.

Perrott, R.H., and Orr, R., A parallel programming language for SIMD and MIMD configurations, in preparation.

Welsh, J., and Bustard, D., 1979, Pascal Plus - Another language for Modular Multiprogramming, Software - Practice and Experience, 9:947-957.

Wirth, N., 1977, Modula: a language for Modular Multiprogramming, Software - Practice and Experience, 9:3-35.

III. SUPERCOMPUTER APPLICATIONS

VECTORISATION TECHNIQUES AND DYNAMIC ELECTRON CORRELATIONS †

F. Brosens* and J. T. Devreese**

Department of Physics
Universitaire Instelling Antwerpen (UIA)
Universiteitsplein 1
B-2610 Antwerpen (Belgium)

Abstract
Some of the vectorisation techniques are discussed, which are of major importance in the development of a numerical procedure to solve the "time-dependent Hartree-Fock" (TDHF) equation for the homogeneous electron gas. This equation accounts for the dynamical exchange effects in the dielectric function. Mathematically, the problem to be solved amounts to the solution of an integral equation in two variables with a singular kernel. The general outline of the analytical and numerical techniques, as well as the physical aspects of the problem have been described elsewhere, and are only briefly summarised. The present paper merely deals with the global vectorisation, realised in the program, and discusses some examples in quite some detail in order to illustrate the procedures used.

† Work supported by the Supercomputer Project of the NFWO (National Fund for Scientific Research, Belgium).
Partially performed in the framework of the collaboration between the Management Unit of the Maritime Environment (MUMM) of the North Sea (Ministry of Public Health and Environment) and the "ALPHA"-Supercomputer-front-end project (NFWO–UIA).

* Senior Research Associate of the National Fund for Scientific Research (Belgium).

** And: Rijksuniversitair Centrum Antwerpen, Groenenborgerlaan 171, B-2020 Antwerpen (Belgium);
and: University of Technology, Eindhoven (The Netherlands).

1. INTRODUCTION

Dynamical exchange effects in the dielectric function of the electron gas are described by the so-called *"time-dependent Hartree-Fock"* (TDHF) equation. The physical background of this equation is sketched in Appendix B, which closely follows the lines and the notations of Ref. [1], where the main results of the numerical solution are discussed, and where we also discuss the relation between the numerical solution and a previous variational approach, to which the present authors contributed [2].

The present paper deals with some vectorisation aspects in the solution of the TDHF-equation, which for a homogeneous electron gas can be written (see Eq. (B.19) in Appendix B) in the form:

$$b_{\vec{k},\nu}(\vec{\xi}) = 1 - \frac{2\pi}{C}\int_S d^2\xi'\left(\frac{\xi'_\perp j_{\vec{\xi}}(-\vec{\xi}'-\vec{k})b^*_{\vec{k},-\nu}(\vec{\xi}')}{\left[(-\nu)+-\Lambda_{\vec{k}}(\vec{\xi}')\right]^*} + \frac{\xi'_\perp j_{\vec{\xi}}(\vec{\xi}')b_{\vec{k},\nu}(\vec{\xi}')}{\nu+-\Lambda_{\vec{k}}(\vec{\xi}')}\right) \tag{1.1}$$

with

$$j_{\vec{\xi}}(\vec{\xi}') = \frac{1}{\sqrt{\left[(\xi_z-\xi'_z)^2+\xi_\perp^2+\xi_\perp'^2\right]^2-4\xi_\perp^2\xi_\perp'^2}} \tag{1.2.a}$$

$$\Lambda_{\vec{k}}(\vec{\xi}) = \frac{k^2}{2} + \vec{\xi}.\vec{k} - \frac{2\pi}{C}\left[T(|\vec{\xi}+\vec{k}|) - T(\xi)\right] \tag{1.2.b}$$

$$T(\xi) = \frac{1-\xi^2}{2\xi}\log\left|\frac{1+\xi}{1-\xi}\right| \tag{1.2.c}$$

where $b_{\vec{k},\nu}(\vec{\xi})$ is related to the Wigner distribution function, but is a smooth function in the two-dimensional domain S which is defined by:

$$\vec{\xi} \in S \quad\Longleftrightarrow\quad |\vec{\xi}| < 1 \ \text{ and } \ |\vec{\xi}+\vec{k}| > 1 \tag{1.3}$$

In this integration domain, $\xi_\perp$ and ξ_z denote the components of $\vec{\xi}$ perpendicular resp. parallel to $\vec{k}$. The integration domain is thus a semicircle of unit radius centered at the origin, of which in the case $k < 2$ the overlap region with a semicircle centered around $\xi_z = -k$ has to be substracted.

From this function $b_{\vec{k},\nu}(\vec{\xi})$, the dielectric function $\epsilon(q,\omega)$ has to be determined:

$$\epsilon(q,\omega) = 1 - \frac{2}{k^2}\frac{2\pi}{C}\int_S d^2\xi'\left(\frac{\xi'_\perp b^*_{\vec{k},-\nu}(\vec{\xi}')}{\left[(-\nu)+-\Lambda_{\vec{k}}(\vec{\xi}')\right]^*} + \frac{\xi'_\perp b_{\vec{k},\nu}(\vec{\xi}')}{\nu+-\Lambda_{\vec{k}}(\vec{\xi}')}\right) \tag{1.4}$$

where $\vec{q}$ and ω are the wave vector and the frequency of an externally applied perturbation. Using the appropriate Fermi-units (nl. the Fermi wave vector k_F and the Fermi energy E_F, these physical quantities are expressed in the dimensionless variables $\vec{k}$ and ν, whereas C is a parameter related to the electron density n of the system:

$$\vec{k} = \frac{\vec{q}}{k_F} \quad ; \quad \nu = \frac{\hbar\omega}{2E_F} \quad ; \quad \xi = \frac{\vec{p}}{\hbar k_F} \quad ; \quad C \equiv 2\pi^2 \left(\frac{9\pi}{4}\right)^{1/3} \frac{1}{r_s} \tag{1.5}$$

The details are given in Appendix B.

In the next section we discuss the numerical procedure, used to solve this equation, with emphasis on the vectorisation aspects. In Appendix A, some considerations about devectorisation are given, which are of some relevance for developping and debugging purposes. Finally, in Appendix B we sketch the physical context of the TDHF-equation.

2. NUMERICAL SOLUTION OF THE TDHF-EQUATION

In Eq. (1.1), the TDHF equation was expressed in terms of a function $b_{\vec{k},\nu}(\vec{\xi})$, which by construction does not show the resonant behavior of the Wigner distribution function in the single-particle continuum.

The vector $\vec{\xi}$ is a two-dimensional vector with components ξ_z and $\xi_\perp$, where the z-component is in the direction of the wave vector $\vec{k}$ under consideration. Because of the smooth behavior of the function $b_{\vec{k},\nu}(\vec{\xi})$, sufficiently small subdomains S_J in the integration domain S can be defined, in which the unknown function can be replaced by its average in these subdomains:

$$b_{\vec{k},\nu}(J) = \frac{1}{S_J} \int_{S_J} d^2\xi b_{\vec{k},\nu}(\vec{\xi}) \tag{2.1}$$

This allows for the discretization of the integral equation (1.1) and the expression (1.4) for the dielectric function:

$$b_{\vec{k},\nu}(I) = 1 - \frac{2\pi}{C} \sum_J \left[F_{\vec{k},\nu}(I,J) b_{\vec{k}\,\nu}(J) + G^*_{\vec{k},\nu}(I,J) b^*_{\vec{k},-\nu}(J)\right] \tag{2.2}$$

$$\epsilon(q,\omega) = 1 - \frac{2}{k^2}\frac{2\pi}{C} \sum_J \left[H_{\vec{k},\nu}(J) b_{\vec{k}\,\nu}(J) + H^*_{\vec{k},\nu}(J) b^*_{\vec{k},-\nu}(J)\right] \tag{2.3}$$

where:

$$F_{\vec{k},\nu}(I,J) = \int_{S_J} d^2\xi \frac{\xi_\perp}{\nu + -\Lambda_{\vec{k}}(\vec{\xi})} \frac{1}{S_I} \int_{S_I} d^2\xi' j_{\vec{\xi}'}(\vec{\xi}) \tag{2.4}$$

$$G_{\vec{k},\nu}(I,J) = \int_{S_J} d^2\xi \frac{\xi_\perp}{\nu + -\Lambda_{\vec{k}}(\vec{\xi})} \frac{1}{S_I} \int_{S_I} d^2\xi' j_{\vec{\xi}'}(\vec{\xi} + \vec{k}) \tag{2.5}$$

$$H_{\vec{k},\nu}(J) = \int_{S_J} d^2\xi \frac{\xi_\perp}{\nu + -\Lambda_{\vec{k}}(\vec{\xi})} \tag{2.6}$$

and with $j_{\vec{\xi}}(\vec{\xi}')$ given by Eq. (1.2).

Since (2.2) is a linear set of equations in the unknowns $b_{\vec{k},\nu}(I)$, standard techniques are available to find the solution, once the matrixelements $F_{\vec{k},\nu}(I,J)$ and $G_{\vec{k},\nu}(I,J)$

are obtained. But these matrixelements require the evaluation of a fourfold integral over the chosen subdomains. To keep the computation time within reasonable limits, a judicious choice of the shape of the subdomains and of the quadrature techniques is in order.

In practice, the subdomains S_J were defined in the following way. The radius of the total integration domain S was subdivided in partitions of equal magnitude. The obtained shells were in turn subdivided with respect to the polar angle, in subdomains of equal surface. For N radial distributions, this leads to N^2 subdomains, and a table can be constructed with the N^2 sets of four data: ξ_J^{min}, ξ_J^{max}, θ_J^{min} and θ_J^{max}, which characterize each subdomain S_J.*

The shape of the subdomains is chosen as described above, because refinements are extremely simple and do not alter the shape. Furthermore, the integration with respect to $\vec{\xi}'$ in (2.4) and (2.5) is relatively easy in these subdomains. It even turns out that these integrals can completely be performed in terms of elliptic functions. But a faster numerical procedure results if only the angular integration is done analytically, leaving the radial integration for numerical evaluation.

Having performed the integration with respect to $\vec{\xi}'$ in (2.4) and (2.5), the three expressions (2.4)–(2.6) have the same structure, which we abbreviate as:

$$\chi_{\vec{k},\nu}^{[f]}(J) = \int_{S_J} d^2\xi \frac{f(\vec{\xi})\xi_\perp}{\nu^+ - \Lambda_{\vec{k}}(\vec{\xi})} \tag{2.7.a}$$

where $f(\vec{\xi})$ is a real function, appropriately chosen for the three cases (2.4)–(2.6):

$$f(\vec{\xi}) = \begin{cases} \frac{1}{S_I}\int_{S_I} d^2\xi' j_{\vec{\xi}'}(\vec{\xi}) & \text{for Eq. (2.4);} \\ \frac{1}{S_I}\int_{S_I} d^2\xi' j_{\vec{\xi}'}(\vec{\xi}+\vec{k}) & \text{for Eq. (2.5);} \\ 1 & \text{for Eq. (2.6).} \end{cases} \tag{2.7.b}$$

Since a large number of integrals of the type (2.7) is to be evaluated, efficient and easily vectorizable procedures are necessary. Since ν^+ is defined as $\nu + i\delta$, with δ an infinitesimal positive constant, the relation

$$\lim_{\delta\to 0} \Im \int_{x_1}^{x_2} dx \frac{g(x)}{x-a-i\delta} = \pi g(a)\delta(a) \quad \text{if} \quad x_1 < a < x_2 \tag{2.8}$$

can be used to eliminate the angular integral in the determination of the imaginary part of (2.7):

$$\Im\chi_{\vec{k},\nu}^{[f]}(J) = \pi \int_{\xi_J^{min}}^{\xi_J^{max}} \xi d\xi \int_{\theta_J^{min}}^{\theta_J^{max}} d\theta f(\vec{\xi})\delta(\nu - \Lambda_{\vec{k}}(\vec{\xi}))\Theta(\Lambda_{\vec{k}}(\xi,\theta_J^{min} - \nu)\Theta(\nu - \Lambda_{\vec{k}}(\xi,\theta_J^{max})) \tag{2.9}$$

where $\Theta(x) = \begin{cases} 1 & \text{for } x > 0; \\ 0 & \text{otherwise.} \end{cases}$

* The case $k < 2$ introduces some minor complications [see Eq. (1.3)], which in practice result in some extra bookkeeping: some of the subdomains do not contribute, and are eliminated from the table, whereas the subdomains intersected by $|\vec{\xi}+\vec{k}| = 1$ require some extra attention. However, these special cases do not influence the concept of the algorithm, and therefore we skip the discussion of these purely technical details.

This equation requires first the determination of the minimum and maximum values ν_J^{min} and ν_J^{max} of ν, for which the δ-function can give a contribution:

$$\left.\begin{aligned} \nu_J^{min} = & \quad \text{Min}\left[\Lambda_{\vec{k}}(\vec{\xi})\right] \\ \nu_J^{max} = & \quad \text{Max}\left[\Lambda_{\vec{k}}(\vec{\xi})\right] \end{aligned}\right\} \quad \text{with} \quad \vec{\xi} \in S_J \tag{2.10}$$

From the analytical expression (1.2.b) for $\Lambda_{\vec{k}}(\vec{\xi})$, one readily checks its monotonical decrease for increasing ξ, which enormously simplifies the numerical determination of ν_J^{min} and ν_J^{max}. Futhermore, this monotonical behavior implies that only one polar angle exists, for which $\nu_J^{min} < \nu < \nu_J^{max}$, satisfying the condition imposed by the δ-function in (2.9):

$$\Im\chi_{\vec{k},\nu}^{[f]}(J) = \pi \int_{\xi_J^{min}}^{\xi_J^{max}} \xi d\xi \frac{f(\vec{\xi})}{|\frac{d}{d\theta}\Lambda_{\vec{k}}(\vec{\xi})|} \quad \text{for} \quad \theta = \theta_0 \quad \text{and for} \quad \nu_J^{min} < \nu < \nu_J^{max} \tag{2.11}$$

and zero elsewhere. In this expression, θ_0 denotes the angle (dependent on $\vec{k}$,ν and $\vec{\xi}$), for which the following condition holds:

$$\Lambda_{\vec{k}}(\xi, \theta_0) = \nu \quad \text{for} \quad \nu_J^{min} < \nu < \nu_J^{max} \tag{2.12}$$

For any given value of ν, ξ and k, the determination of the corresponding θ_0 is an easy numerical task, and the derivative in (2.11) can be calculated analytically. For practical purposes, we have chosen a sufficiently dense mesh of N_ν equidistant frequencies ν_{J,i_ν} in each subdomain S_J, with N_ν in the order 10 to 20.

$$\nu_{J,i_\nu} = \nu_J^{min} + \frac{i_\nu - 1}{N_\nu - 1}\left(\nu_J^{max} - \nu_J^{min}\right) \quad i_\nu = 1, 2, \cdots, N_\nu \tag{2.13}$$

But the evaluation of the integral (2.11) then first requires the determination of the minimum and maximum value of ξ in the range $[\xi_J^{min}, \xi_J^{max}]$ for which (2.12) can be satisfied. This evaluation has to be performed numerically. Denoting these values by ξ_{J,i_ν}^{min} and ξ_{J,i_ν}^{max} in each subdomain S_J for each frequency ν_{J,i_ν}, a mesh of N_ξ points has to be chosen:

$$\xi_{J,i_\nu}^{min} \leq \xi_{J,i_\nu,i_\xi} \leq \xi_{J,i_\nu}^{max} \quad ; \quad i_\xi = 1, 2, \cdots, N_\xi \tag{2.14}$$

according to the quadrature scheme one likes to use. We compared several integration routines, ranging from the trapezium rule to Gauss-Legendre and Chebychev quadratures. But the smoothness of the integrand combined with the smallness of the integration domain, have as a consequence that all these procedures are in essence equivalent. In practice, the Gauss-Legendre scheme with $N_\xi \approx 10$ is quite accurate.

Obviously, this procedure is highly vectorizable. With only 10 subdomains, 10 frequencies per subdomain, and 10 mesh points ξ for each frequency per subdomain, the data ξ_{J,i_ν,i_ξ} with their corresponding angles θ_{J,i_ν,i_ξ} and weighting factors w_{J,i_ν,i_ξ} already constitute vectors of length 1000. Constructing the corresponding function values $f(\vec{\xi})$ as a vector function on a Cyber-205 thus reduces the evaluation of (2.9) to a fully vectorized operation. But also the result $\Im\chi_{\vec{k},\nu}^{[f]}(J)$ is a vector in the interval counters J and in the frequencies ν_{J,i_ν} as given by (2.13).

It is then important that also the real part of this function via the Kramers-Kronig relation:

$$\Re\chi_{\vec{k},\nu}^{[f]}(J) = \frac{1}{\pi} P \int_{\nu_J^{min}}^{\nu_J^{max}} d\lambda \frac{\Im\chi_{\vec{k},\nu}^{[f]}(J)}{\lambda - \nu} \tag{2.15}$$

can be obtained in vector-mode. (P denotes that the principal value of the integral is considered.) In the actual calculation, this was realized by expressing $\Im\chi^{[f]}_{\vec{k},\nu}(J)$ in cubic spline polynomials for the chosen frequency mesh (2.13).

In general, the cubic spline interpolation in its scalar version is a well known procedure to approximate a function $f(x)$ in an interval $x_i \leq x \leq x_{i+1}$ as:

$$f(x) = a_i + b_i(x - x_i) + c_i(x - x_i)^2 + d_i(x - x_i)^3 \quad ; \quad x_i \leq x \leq x_{i+1}$$

In particular, for the function $\Im\chi^{[f]}_{\vec{k},\nu}(J)$ of Eq. (2.15) as specified in (2.7.b) and (2.4), this piecewise polynomial fitting thus would become (for a fixed value of $\vec{k}$):

$$\Im F_{\vec{k},\nu}(I,J) = A^{[F_{\vec{k}}]}_{I,J,i_\nu} + B^{[F_{\vec{k}}]}_{I,J,i_\nu}(\nu - \nu_{J,i_\nu}) + C^{[F_{\vec{k}}]}_{I,J,i_\nu}(\nu - \nu_{J,i_\nu})^2 + D^{[F_{\vec{k}}]}_{I,J,i_\nu}(\nu - \nu_{J,i_\nu})^3$$
$$\text{for } \nu_{J,i_\nu} \leq \nu \leq \nu_{J,i_\nu+1} \tag{2.16}$$

Similar expansions have to be considered for the functions $\Im G_{\vec{k},\nu}(I,J)$ and $\Im H_{\vec{k},\nu}(J)$.

As an example of the global vectorisation process used, we study the cubic spline vectorisation in some more detail. As a starting point, consider the scalar cubic spline smoothing of Ref. [3], which has the following FORTRAN implementation:

```
      SUBROUTINE SMOOTH(N1,N2,X,Y,DY,S,A,B,C,D)
C* GIVEN SMOOTHING CONDITION S AND GIVEN DATA Y(N) WITH ERROR
C* DY(N) IN POINTS X(N) WITH  N=N1,...,N2   (N2 .LE. 299)
C*
C* CALCULATE COEFFICIENTS A(N=N1,...,N2), B(N=N1,...,N2-1),
C*                        C(N=N1,...,N2), D(N=N1,...,N2-1),
C* SUCH THAT Y(XX) FOR X(N) .LE. XX .LT. X(N+1) IS APPROXIMATED
C* AS          Y(XX)=A(N)+B(N)*H+C(N)*H**2+D(N)*H**3
C*             WITH H=XX-X(N)
C*
      REAL  R(300),R1(300),R2(300),T(300),T1(300),U(300),V(300)
      DIMENSION X(N2),Y(N2),DY(N2),A(N2),B(N2),C(N2),D(N2)
C* INITIALISATION
      M1=N1-1
      M2=N2+1
      R(M1+1)=0.
      R(N1+1)=0.
      R1(N2+1)=0.
      R2(N2+1)=0.
      R2(M2+1)=0.
      U(M1+1)=0.
      U(N1+1)=0.
      U(N2+1)=0.
      U(M2+1)=0.
C* FIRST ITERATION
      P=0.
      M1=N1+1
      M2=N2-1
```

```
      H=X(M1)-X(N1)
      F=Y(M1)-Y(N1)
      F=F/H
      DO 2 I=M1,M2
      G=H
      H=X(I+1)-X(I)
      E=F
      F=Y(I+1)-Y(I)
      F=F/H
      A(I)=F-E
      T(I)=2.*(G+H)/3.
      T1(I)=H/3.
      R2(I+1)=DY(I-1)/G
      R(I+1)=DY(I+1)/H
      R1(I+1)=-DY(I)/G-DY(I)/H
2     CONTINUE
      DO 3 I=M1,M2
      B(I)=R(I+1)*R(I+1)+R1(I+1)*R1(I+1)+R2(I+1)*R2(I+1)
      C(I)=R(I+1)*R1(I+2)+R1(I+1)*R2(I+2)
      D(I)=R(I+1)*R2(I+3)
3     CONTINUE
      F2=-S
C* NEXT ITERATION
50    DO 4 I=M1,M2
      R1(I)=F*R(I)
      R2(I-1)=G*R(I-1)
      R(I+1)=1/( P*B(I)+T(I)-F*R1(I)-G*R2(I-1) )
      U(I+1)=A(I)-R1(I)*U(I)-R2(I-1)*U(I-1)
      F=P*C(I)+T1(I)-H*R1(I)
      G=H
      H=D(I)*P
4     CONTINUE
      DO 5 J=M1,M2
      I=M1+M2-J
      U(I+1)=R(I+1)*U(I+1)-R1(I+1)*U(I+2)-R2(I+1)*U(I+3)
5     CONTINUE
      E=0.
      H=0.
      DO 6 I=N1,M2
      G=H
      H= ( U(I+2)-U(I+1) )/( X(I+1)-X(I) )
      V(I)=(H-G)*DY(I)*DY(I)
      E=E+V(I)*(H-G)
6     CONTINUE
      V(N2)=-H*DY(N2)*DY(N2)
      G=V(N2)
      E=E-G*H
      G=F2
      F2=E*P*P
      IF ( F2.GE.S.OR.F2.LE.G) GO TO 100
      F=0.
```

```
      H=(V(M1)-V(N1))/(X(M1)-X(N1))
      DO 7 I=M1,N2
      G=H
      H=(V(I+1)-V(I))/(X(I+1)-X(I))
      G=H-G-R1(I)*R(I)-R2(I-1)*R(I-1)
      F=F+G*R(I+1)*G
      R(I+1)=G
7    CONTINUE
      H=E-P*F
      IF (H.LE.0) GO TO 100
      P=P+(S-F2)/(H*(SQRT(S/E)+P))
      GO TO 50
C* FINAL RESULT
100   DO 101 I=N1,N2
      A(I)=Y(I)-P*V(I)
      C(I)=U(I+1)
101   CONTINUE
      DO 102 I=N1,M2
      H=X(I+1)-X(I)
      D(I)=(C(I+1)-C(I))/(3.*H)
      B(I)=(A(I+1)-A(I))/H-H*(H*D(I)+C(I))
102   CONTINUE
      RETURN
      END
```

As is discussed in Ref. [3], this smoothing procedure is recursive in nature. If limited to a single function at once, its vectorisation is therefore necessarily rather unsatisfactory. However, in the case (2.15) under consideration here, we are interested in the approximation of a function for each subdomain, i. e. we can consider **a vector of functions**, for which an optimal vectorisation can be realised. This means that the expansion coefficients $A^{[F_{\vec{\xi}}]}_{I,J,i_\nu}$, $B^{[F_{\vec{\xi}}]}_{I,J,i_\nu}$, $C^{[F_{\vec{\xi}}]}_{I,J,i_\nu}$ and $D^{[F_{\vec{\xi}}]}_{I,J,i_\nu}$ [and similarly for the coefficients related to $\Im G_{\vec{\xi},\nu}(I,J)$ and $\Im H_{\vec{\xi},\nu}(J)$] can be considered as **vectors** in the indices I, J for the subdomains, and as **arrays** for the splining in the indices i_ν for the frequencies considered. (In practice, one thus defines arrays of vectors with a vector-length of order 20 000, which means that the vectorisation is very efficient.) The vectorized implementation of the cubic spline interpolation (with the approximated function forced through the input-data) which we actually used is given below, and extensively takes profit of the **vector variables**, which are available on CYBER-205.

```
      SUBROUTINE VSPLINE(N1,N2,X,Y,B,C,D)
C* GIVEN DATA Y(N) IN POINTS X(N) WITH  N=N1,...,N2 (N2 .LE. 29)
C*
C* CALCULATE COEFFICIENTS B(N=N1,...,N2-1),
C*                        C(N=N1,...,N2), D(N=N1,...,N2-1),
C* SUCH THAT Y(XX) FOR X(N) .LE. XX .LT. X(N+1) IS APPROXIMATED
C* AS       Y(XX)=Y(N)+B(N)*H+C(N)*H**2+D(N)*H**3
C*          WITH H=XX-X(N)
C-------------------------------------
C X, Y, B, C, D ARE DESCRIPTOR-ARRAYS; I.E. WE CONSIDER NDESC
C FUNCTIONS SIMULTANEOUSLY, WHERE NDESC IS THE LENGTH OF THE
```

```
C DESCRIPTOR. THE NUMBER OF GIVEN FUNCTION VALUES IS ASSUMED TO
C BE EQUAL FOR EACH FUNCTION.
C-------------------------------------------
      DESCRIPTOR T(30),T1(30),Q(30),V(30),H(30)
      DESCRIPTOR X(N2),Y(N2),B(N2),C(N2),D(N2)
      NFUN=Q8SLEN(X(1))
      DO 100 I=N1,N2
      ASSIGN T(I),.DYN.NFUN
      ASSIGN T1(I),.DYN.NFUN
      ASSIGN Q(I),.DYN.NFUN
      ASSIGN V(I),.DYN.NFUN
      ASSIGN H(I),.DYN.NFUN
 100  CONTINUE
       DO 999 I=N1,N2
       B(I)=0.
       C(I)=0.
       D(I)=0.
       T(I)=0.
       T1(I)=0.
       Q(I)=0.
       V(I)=0.
       H(I)=0.
 999  CONTINUE
      M1=N1+1
      M2=N2-1
      DO 1 I=N1,M2
 1    H(I)=X(I+1)-X(I)
      DO 2 I=M1,M2
      V(I)=Y(I-1)/H(I-1)+Y(I+1)/H(I)-Y(I)*(1./H(I-1)+1./H(I))
      T(I)=(2./3.)*(H(I-1)+H(I))
 2    T1(I)=H(I)/3.
      B(N1)=T1(M1)/T(M1)
      M3=N2-2
      IF (M3.GE.M1) THEN
      DO 3 I=M1,M2-1
 3    B(I)=T1(I+1)/(T(I+1)-T1(I)*B(I-1))
      ENDIF
      Q(N1)=V(M1)/T(M1)
      DO 4 I=M1,M2
 4    Q(I)=(V(I+1)-T1(I)*Q(I-1))/(T(I+1)-T1(I)*B(I-1))
      C(M2)=Q(M3)
      DO 5 I=M3,N1,-1
 5    C(I+1)=Q(I)-B(I)*C(I+2)
      DO 10 I=N1,M2
      D(I)=(C(I+1)-C(I))/(3.*H(I))
 10   B(I)=(Y(I+1)-Y(I))/H(I)-H(I)*(H(I)*D(I)+C(I))
      RETURN
      END
```

For a function of the form:

$$f(x) = f_i + b_i(x - x_i) + c_i(x - x_i)^2 + d_i(x - x_i)^3 \quad ; \quad x_i \leq x \leq x_{i+1}$$

the principle value integral from the Kramers-Kronig relation (2.15) for the real part of these functions becomes:

$$P\int_{x_i}^{x_{i+1}} dx\frac{f(x)}{x-p} = f_i S_0(x_i,p) + b_i S_1(x_i,p) + c_i S_2(x_i,p) + d_i S_3(x_i,p)$$

with

$$S_0(x_i,p) = \log\left|\frac{x_{i+1}-p}{x_i-p}\right|$$

$$S_{n>0}(x_i,p) = \frac{1}{n}(x_{i+1}-x_i)^n + (p-x_i)S_{n-1}(x_i,p)$$

For the functions $\Im F_{\vec{k},\nu}(I,J)$, $\Im G_{\vec{k},\nu}(I,J)$ and $\Im H_{\vec{k},\nu}(J)$ considered here, the expansion coefficients $A_{I,J,i_\nu}^{[F_{\vec{k}}]}, \ldots$ are vector output arguments of the interpolation procedure, as discussed above. The determination of the real part from the principal value integral is thus very easy to vectorize, since the functions involved are intrinsic vector functions on CYBER-205.

Having thus obtained the matrix elements $F_{\vec{k},\nu}(I,J)$ and $G_{\vec{k},\nu}(I,J)$, defined in (2.4) and (2.5), and the vector $H_{\vec{k},\nu}(J)$ in (2.6) determining the dielectric function, the evaluation of the unknowns $b_{\vec{k},\nu}(J)$ from (2.2) can be obtained with standard matrix algebra packages.

For a discussion of the results, we refer to Ref. [1]. The main purpose of the present paper is to show the basic idea of the **global vectorisation**, realized in the numerical solution of the TDHF equation, and which was essentially performed by using large vectors as arguments of subroutines, of which almost standard scalar versions exist. The cubic-spline smoothing given above is a typical example of this approach.

APPENDIX A: "DEVECTORISATION"

Vectorisation is obviously one of the major techniques for speeding up many application programs. But the vector-manuals often stress the hardware components and the language aspects at the machine level. Far less attention is payed to the vector extensions in the Fortran language, as provided by ETA-10 and CYBER-205. Nevertheless, these extensions are very interesting for realising 'global' vectorisation, in contrast to the 'local' vectorisation (which is by definition the level of vectorisation by auto-vectorisers and programs like VAST and KAP).

Because of the rather limited range of computers with vector extensions at the level of the standard programming language, many users are hesitating to fully explore the vectorisation technique. A second threshold is the lack of interactive and user-friendly testing and debugging techniques on most supercomputers. Batch-oriented machines are not very appropriate for the development of new algorithms for scientific applications.

In order to overcome these hindrances for a wider use of the vector facilities on

CYBER-205, the devectorisation tool **VECTORA** has been developed. **VECTORA*** is a program, written in scalar FORTRAN (essentially FORTRAN-77), to translate source codes in Vector-FORTRAN into source codes in scalar FORTRAN.

- The input for **VECTORA** is a FORTRAN-77 source code with the vector extensions of CYBER-205 and ETA-10 (i.e. FORTRAN-200). These vector features include vector variables ('DESCRIPTORS'), bit variables and vector functions, which are basic tools for 'global' vectorisation. Therefore, vectorisation is easily realised at a level beyond the capacities of any 'auto-vectoriser'.

- The output of **VECTORA** is again a FORTRAN-77 source code, to be linked with a library provided by **VECTORA**, and which includes **scalar emulations** for the typical vector functions of FORTRAN-200.

- Usage: **VECTORA** allows to execute vector codes on scalar machines (of course with limitations in memory capacity and CPU-time).

For testing and debugging purposes, the direct use of a supercomputer can thus be avoided, and diagnostics are provided in the common environment of the user, with the possibility of interactive testing. Furthermore, the software packages and facilities which are familiar to the programmer on his scalar machine, remain at his disposal.

In order to allow the interactive testing and debugging of vector-codes in the development stage of a program –especially for unexperienced users–, VECTORA allows to overcome many of the difficulties encountered in using a CYBER-205 (e.g. the BATCH-environment, the poor debugging facilities, the lack of user-friendly vectorisation tools). It allows program development directly in vector code, with the standard interactive facilities of a scalar machine. One of the main advantages is the fact that it also translates VECTORS (≢ ARRAYS) used as arguments, which is in practice very hard to realise manually.

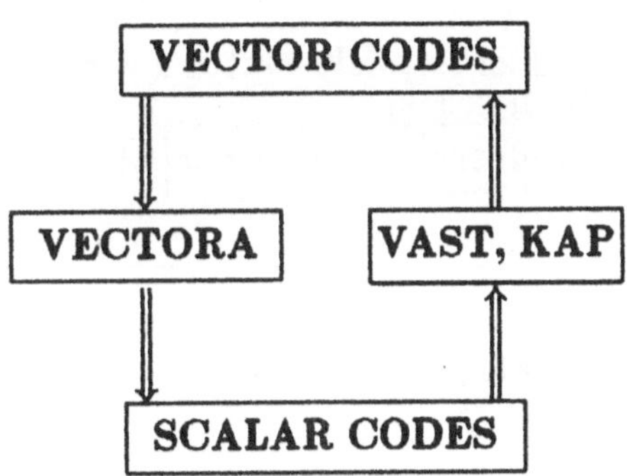

The translation of vector codes into scalar codes includes:
- BIT variables (not constants);
- DESCRIPTOR variables;
- WHERE blocks;
- ASSIGN descriptor (also .DYN.);
- implicit and explicit vector references;

* Available form Control Data–The Netherlands

– user-defined vector functions;
– predefined functions (Q8... etc.);
– ...

As an illustration, we show some schematic examples:

Explicit vector reference:

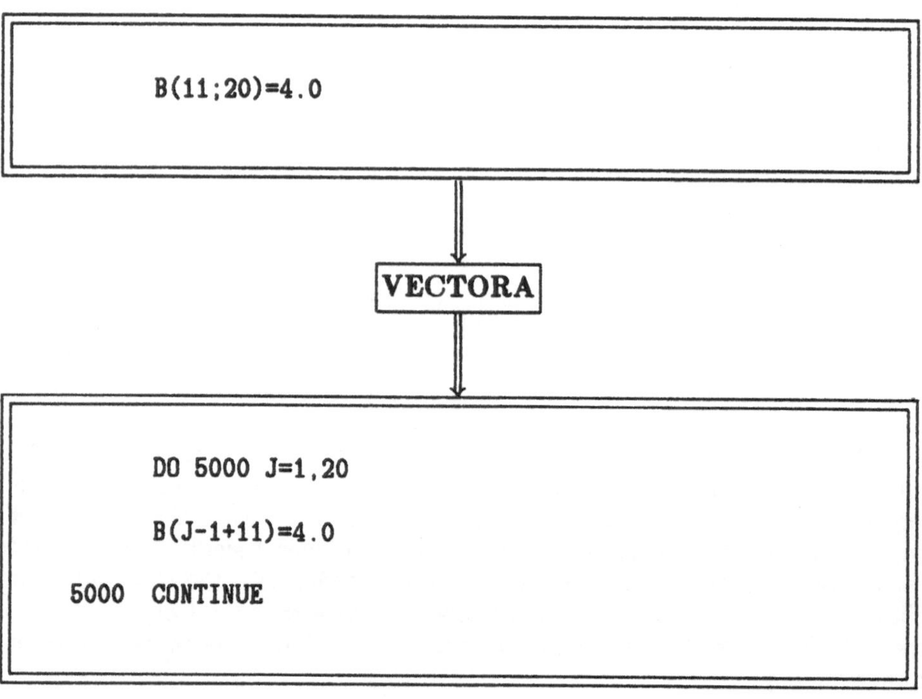
B(11;20)=4.0
VECTORA
DO 5000 J=1,20
B(J-1+11)=4.0
5000 CONTINUE

The application with an **Implicit vector reference** introduces a variable DOUBLE PRECISION DESClnm
with contents: `Start address` `Type` `Length` of the vector.

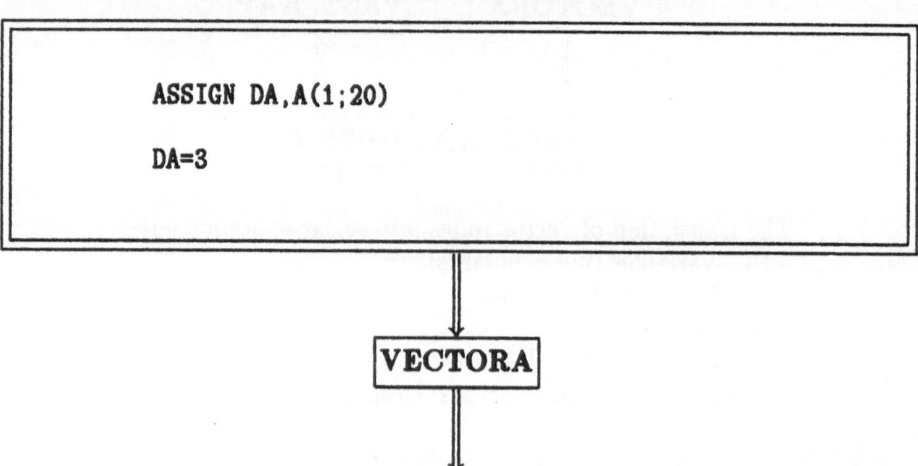
ASSIGN DA,A(1;20)
DA=3
VECTORA

```
      CALL DESCASS(DESC001,LOCA(A(1)),DESCTYP(A(1)),20)

      DO 5000 J=1,Q8SLEN(DESC001)

      DA=3

5000  CALL DESCOFV(LOCA(DA),DESCTYP(DA),DESC001,J)
```

A third schematic example includes a **WHERE statement with DESCRIPTORS:**

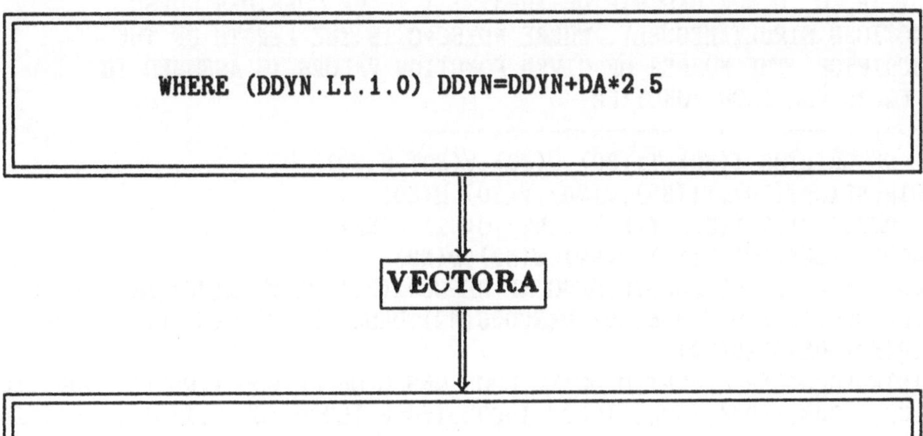

WHERE (DDYN.LT.1.0) DDYN=DDYN+DA*2.5
VECTORA


```
      DO 5000 J=1,Q8SLEN(DESC004)

      CALL DESCTOV(LOCA(DDYN),DESCTYP(DDYN),DESC004,J)

      IF (DDYN.LT.1.0) THEN

      CALL DESCTOV(LOCA(DA),DESCTYP(DA),DESC001,J)

      CALL DESCTOV(LOCA(DDYN),DESCTYP(DDYN),DESC004,J)

      DDYN=DDYN+DA*2.5

      CALL DESCOFV(LOCA(DDYN),DESCTYP(DDYN),DESC004,J)

      ENDIF

5000  CONTINUE
```

As a final example, we show the translation of the cubic spline code, discussed above, wich illustrates the power of the routine as a tool for realising global vectorisation, which is extremely useful in the development and debugging stage of a program.

```
C  TRANSLATED VERSION OF THE PROGRAM BY THE EMULATOR
C        ( PRELIMINARY EMULATOR VERSION )

C*        SUBROUTINE VSPLINE(N1,N2,X,Y,B,C,D)
      SUBROUTINEVSPLINE(N1,N2,X,DESC006,Y,DESC007,B,DESC008,C,DESC009,D,
     +DESC010)
C*C* GIVEN DATA Y(N) IN POINTS X(N) WITH  N=N1,...,N2 (N2 .LE. 29)
C*C*
C*C* CALCULATE COEFFICIENTS B(N=N1,...,N2-1),
C*C*                        C(N=N1,...,N2), D(N=N1,...,N2-1),
C*C* SUCH THAT Y(XX) FOR X(N) .LE. XX .LT. X(N+1) IS APPROXIMATED
C*C* AS         Y(XX)=Y(N)+B(N)*H+C(N)*H**2+D(N)*H**3
C*C*            WITH H=XX-X(N)
C---------------------------------------
C*C X, Y, B, C, D ARE DESCRIPTOR-ARRAYS; I.E. WE CONSIDER NDESC
C*C FUNCTIONS SIMULTANEOUSLY, WHERE NDESC*C IS THE LENGTH OF THE
C*C DESCRIPTOR. THE NUMBER OF GIVEN FUNCTION VALUES IS ASSUMED TO
C*C BE EQUAL FOR EACH FUNCTION.
C---------------------------------------
C*        DESCRIPTOR T(30),T1(30),Q(30),V(30),H(30)
      DIMENSIONT(30),T1(30),Q(30),V(30),H(30)
C*        DESCRIPTOR X(N2),Y(N2),B(N2),C(N2),D(N2)
      DIMENSIONX(N2),Y(N2),B(N2),C(N2),D(N2)
      DOUBLEPRECISIONDESCDYN,DESCTMP,DESC001(30),DESC002(30),DESC003(30)
     +,DESC004(30),DESC005(30),DESC006(N2),DESC007(N2),DESC008(N2),DESCO
     +09(N2),DESC010(N2)
      INTEGERQ8SLEN,Q8SCNT,DESCTYP,LOCA,Q8SEQ,Q8SGE,Q8SLT,Q8SNE,I5001,I5
     +002,I5003,I5004,I5005,I5006,I5007,I5008,I5009,I5010,I5011,I5012,I5
     +013,I5014,I5015,I5016,I5017,I5018,I5019,I5020
      LOGICALBTOL,LTOB
C*        NFUN=Q8SLEN(X(1))
      NFUN=Q8SLEN(DESC006(1))
      DO 100 I=N1,N2
C*        ASSIGN T(I),.DYN.NFUN
      CALLDYNASSI(DESCDYN,DESC001(I),LOCA(T(I)),DESCTYP(T(I)),NFUN)
C*        ASSIGN T1(I),.DYN.NFUN
      CALLDYNASSI(DESCDYN,DESC002(I),LOCA(T1(I)),DESCTYP(T1(I)),NFUN)
C*        ASSIGN Q(I),.DYN.NFUN
      CALLDYNASSI(DESCDYN,DESC003(I),LOCA(Q(I)),DESCTYP(Q(I)),NFUN)
C*        ASSIGN V(I),.DYN.NFUN
      CALLDYNASSI(DESCDYN,DESC004(I),LOCA(V(I)),DESCTYP(V(I)),NFUN)
C*        ASSIGN H(I),.DYN.NFUN
      CALLDYNASSI(DESCDYN,DESC005(I),LOCA(H(I)),DESCTYP(H(I)),NFUN)
  100 CONTINUE
       DO 999 I=N1,N2
C*         B(I)=0.
      DO5001I5001=1,Q8SLEN(DESC008)
      B(I)=0.
```

```
5001  CALLDESCOFV(LOCA(B(I)),DESCTYP(B(I)),DESC008(I),I5001)
C*       C(I)=0.
      DO5002I5002=1,Q8SLEN(DESC009)
      C(I)=0.
5002  CALLDESCOFV(LOCA(C(I)),DESCTYP(C(I)),DESC009(I),I5002)
C*       D(I)=0.
      DO5003I5003=1,Q8SLEN(DESC010)
      D(I)=0.
5003  CALLDESCOFV(LOCA(D(I)),DESCTYP(D(I)),DESC010(I),I5003)
C*       T(I)=0.
      DO5004I5004=1,Q8SLEN(DESC001)
      T(I)=0.
5004  CALLDESCOFV(LOCA(T(I)),DESCTYP(T(I)),DESC001(I),I5004)
C*       T1(I)=0.
      DO5005I5005=1,Q8SLEN(DESC002)
      T1(I)=0.
5005  CALLDESCOFV(LOCA(T1(I)),DESCTYP(T1(I)),DESC002(I),I5005)
C*       Q(I)=0.
      DO5006I5006=1,Q8SLEN(DESC003)
      Q(I)=0.
5006  CALLDESCOFV(LOCA(Q(I)),DESCTYP(Q(I)),DESC003(I),I5006)
C*       V(I)=0.
      DO5007I5007=1,Q8SLEN(DESC004)
      V(I)=0.
5007  CALLDESCOFV(LOCA(V(I)),DESCTYP(V(I)),DESC004(I),I5007)
C*       H(I)=0.
      DO5008I5008=1,Q8SLEN(DESC005)
      H(I)=0.
5008  CALLDESCOFV(LOCA(H(I)),DESCTYP(H(I)),DESC005(I),I5008)
 999  CONTINUE
      M1=N1+1
      M2=N2-1
      DO 1 I=N1,M2
C* 1    H(I)=X(I+1)-X(I)
      DO5009I5009=1,Q8SLEN(DESC005)
      CALLDESCTOV(LOCA(X(I+1)),DESCTYP(X(I+1)),DESC006(I+1),I5009)
      CALLDESCTOV(LOCA(X(I)),DESCTYP(X(I)),DESC006(I),I5009)
      H(I)=X(I+1)-X(I)
5009  CALLDESCOFV(LOCA(H(I)),DESCTYP(H(I)),DESC005(I),I5009)
1     CONTINUE
      DO 2 I=M1,M2
C*      V(I)=Y(I-1)/H(I-1)+Y(I+1)/H(I)-Y(I)*(1./H(I-1)+1./H(I))
      DO5010I5010=1,Q8SLEN(DESC004)
      CALLDESCTOV(LOCA(H(I-1)),DESCTYP(H(I-1)),DESC005(I-1),I5010)
      CALLDESCTOV(LOCA(H(I)),DESCTYP(H(I)),DESC005(I),I5010)
      CALLDESCTOV(LOCA(Y(I-1)),DESCTYP(Y(I-1)),DESC007(I-1),I5010)
      CALLDESCTOV(LOCA(Y(I+1)),DESCTYP(Y(I+1)),DESC007(I+1),I5010)
      CALLDESCTOV(LOCA(Y(I)),DESCTYP(Y(I)),DESC007(I),I5010)
      V(I)=Y(I-1)/H(I-1)+Y(I+1)/H(I)-Y(I)*(1./H(I-1)+1./H(I))
5010  CALLDESCOFV(LOCA(V(I)),DESCTYP(V(I)),DESC004(I),I5010)
C*      T(I)=(2./3.)*(H(I-1)+H(I))
```

```
      DO5011I5011=1,Q8SLEN(DESC001)
      CALLDESCTOV(LOCA(H(I-1)),DESCTYP(H(I-1)),DESC005(I-1),I5011)
      CALLDESCTOV(LOCA(H(I)),DESCTYP(H(I)),DESC005(I),I5011)
      T(I)=(2./3.)*(H(I-1)+H(I))
5011  CALLDESCOFV(LOCA(T(I)),DESCTYP(T(I)),DESC001(I),I5011)
C* 2    T1(I)=H(I)/3.
      DO5012I5012=1,Q8SLEN(DESC002)
      CALLDESCTOV(LOCA(H(I)),DESCTYP(H(I)),DESC005(I),I5012)
      T1(I)=H(I)/3.
5012  CALLDESCOFV(LOCA(T1(I)),DESCTYP(T1(I)),DESC002(I),I5012)
2     CONTINUE
C*      B(N1)=T1(M1)/T(M1)
      DO5013I5013=1,Q8SLEN(DESC008)
      CALLDESCTOV(LOCA(T(M1)),DESCTYP(T(M1)),DESC001(M1),I5013)
      CALLDESCTOV(LOCA(T1(M1)),DESCTYP(T1(M1)),DESC002(M1),I5013)
      B(N1)=T1(M1)/T(M1)
5013  CALLDESCOFV(LOCA(B(N1)),DESCTYP(B(N1)),DESC008(N1),I5013)
      M3=N2-2
      IF (M3.GE.M1) THEN
      DO 3 I=M1,M2-1
C* 3    B(I)=T1(I+1)/(T(I+1)-T1(I)*B(I-1))
      DO5014I5014=1,Q8SLEN(DESC008)
      CALLDESCTOV(LOCA(T(I+1)),DESCTYP(T(I+1)),DESC001(I+1),I5014)
      CALLDESCTOV(LOCA(T1(I+1)),DESCTYP(T1(I+1)),DESC002(I+1),I5014)
      CALLDESCTOV(LOCA(T1(I)),DESCTYP(T1(I)),DESC002(I),I5014)
      CALLDESCTOV(LOCA(B(I-1)),DESCTYP(B(I-1)),DESC008(I-1),I5014)
      B(I)=T1(I+1)/(T(I+1)-T1(I)*B(I-1))
5014  CALLDESCOFV(LOCA(B(I)),DESCTYP(B(I)),DESC008(I),I5014)
3     CONTINUE
      ENDIF
C*      Q(N1)=V(M1)/T(M1)
      DO5015I5015=1,Q8SLEN(DESC003)
      CALLDESCTOV(LOCA(T(M1)),DESCTYP(T(M1)),DESC001(M1),I5015)
      CALLDESCTOV(LOCA(V(M1)),DESCTYP(V(M1)),DESC004(M1),I5015)
      Q(N1)=V(M1)/T(M1)
5015  CALLDESCOFV(LOCA(Q(N1)),DESCTYP(Q(N1)),DESC003(N1),I5015)
      DO 4 I=M1,M2
C* 4    Q(I)=(V(I+1)-T1(I)*Q(I-1))/(T(I+1)-T1(I)*B(I-1))
      DO5016I5016=1,Q8SLEN(DESC003)
      CALLDESCTOV(LOCA(T(I+1)),DESCTYP(T(I+1)),DESC001(I+1),I5016)
      CALLDESCTOV(LOCA(T1(I)),DESCTYP(T1(I)),DESC002(I),I5016)
      CALLDESCTOV(LOCA(Q(I-1)),DESCTYP(Q(I-1)),DESC003(I-1),I5016)
      CALLDESCTOV(LOCA(V(I+1)),DESCTYP(V(I+1)),DESC004(I+1),I5016)
      CALLDESCTOV(LOCA(B(I-1)),DESCTYP(B(I-1)),DESC008(I-1),I5016)
      Q(I)=(V(I+1)-T1(I)*Q(I-1))/(T(I+1)-T1(I)*B(I-1))
5016  CALLDESCOFV(LOCA(Q(I)),DESCTYP(Q(I)),DESC003(I),I5016)
4     CONTINUE
C*      C(M2)=Q(M3)
      DO5017I5017=1,Q8SLEN(DESC009)
      CALLDESCTOV(LOCA(Q(M3)),DESCTYP(Q(M3)),DESC003(M3),I5017)
      C(M2)=Q(M3)
```

```
5017  CALLDESCOFV(LOCA(C(M2)),DESCTYP(C(M2)),DESC009(M2),I5017)
      DO 5 I=M3,N1,-1
C* 5    C(I+1)=Q(I)-B(I)*C(I+2)
      DO5018I5018=1,Q8SLEN(DESC009)
      CALLDESCTOV(LOCA(Q(I)),DESCTYP(Q(I)),DESC003(I),I5018)
      CALLDESCTOV(LOCA(B(I)),DESCTYP(B(I)),DESC008(I),I5018)
      CALLDESCTOV(LOCA(C(I+2)),DESCTYP(C(I+2)),DESC009(I+2),I5018)
      C(I+1)=Q(I)-B(I)*C(I+2)
5018  CALLDESCOFV(LOCA(C(I+1)),DESCTYP(C(I+1)),DESC009(I+1),I5018)
5     CONTINUE
      DO 10 I=N1,M2
C*      D(I)=(C(I+1)-C(I))/(3.*H(I))
      DO5019I5019=1,Q8SLEN(DESC010)
      CALLDESCTOV(LOCA(H(I)),DESCTYP(H(I)),DESC005(I),I5019)
      CALLDESCTOV(LOCA(C(I+1)),DESCTYP(C(I+1)),DESC009(I+1),I5019)
      CALLDESCTOV(LOCA(C(I)),DESCTYP(C(I)),DESC009(I),I5019)
      D(I)=(C(I+1)-C(I))/(3.*H(I))
5019  CALLDESCOFV(LOCA(D(I)),DESCTYP(D(I)),DESC010(I),I5019)
C* 10   B(I)=(Y(I+1)-Y(I))/H(I)-H(I)*(H(I)*D(I)+C(I))
      DO5020I5020=1,Q8SLEN(DESC008)
      CALLDESCTOV(LOCA(H(I)),DESCTYP(H(I)),DESC005(I),I5020)
      CALLDESCTOV(LOCA(Y(I+1)),DESCTYP(Y(I+1)),DESC007(I+1),I5020)
      CALLDESCTOV(LOCA(Y(I)),DESCTYP(Y(I)),DESC007(I),I5020)
      CALLDESCTOV(LOCA(C(I)),DESCTYP(C(I)),DESC009(I),I5020)
      CALLDESCTOV(LOCA(D(I)),DESCTYP(D(I)),DESC010(I),I5020)
      B(I)=(Y(I+1)-Y(I))/H(I)-H(I)*(H(I)*D(I)+C(I))
5020  CALLDESCOFV(LOCA(B(I)),DESCTYP(B(I)),DESC008(I),I5020)
10    CONTINUE
      CALLDESCFRE(DESCDYN)
      RETURN
      END
```

APPENDIX B: PHYSICAL BACKGROUND OF THE TDHF-EQUATION

In this section, which closely follows Ref. [1], we sketch the physical assumptions underlying the TDHF-equation, which is used for the description of the dynamical exchange effects in the dielectric function of the electron gas. These effects are usually described by the so-called local-field correction $G(q,\omega)$ in the dielectric function:

$$\epsilon(q,\omega) = 1 + \frac{Q_0(q,\omega)}{1 - G(q,\omega)Q_0(q,\omega)} \tag{B.1}$$

In this expression, $Q_0(q,\omega)$ is the well known Lindhard function [4]. The neglect of $G(q,\omega)$ leads to the dielectric function in the "Random-Phase-Approximation" (RPA) [5]. The present authors have contributed [6] in the development of a variational method for including dynamical exchange effects in $G(q,\omega)$.

The RPA only takes the **Hartree** interaction into account in the electron dynamics. This is the basic reason why its remarkable succes is limited to the long-wavelength limit, e.g. in the plasmon dispersion. To account for the experimental plasmon spectrum [7–8] for wave vectors of the order of the Fermi wave vector, exchange and correlation effects (and the interaction with the lattice) are an indispensable ingredient in the dielectric function. The failure of the RPA to describe short-range correlations also leads to rather insatisfactory results with regard to sum rules. E.g. the pair correlation function $g(r)$, as calculated from the RPA dielectric function, becomes negative in the origin [9].

The formal introduction of exchange and correlation effects in the dielectric function by the the local-field correction $G(q,\omega)$, as given in the Eq. (B.1) , has become a rather standard description since the work of Hubbard [10]. (In this work, the frequency dependence was not yet considered.) A large variety of approximations has since then been proposed [11] to improve upon the RPA. Most of these approximations are in essence dealing with the problem of the exchange contributions, since no practical and systematic way seems to be known to account for dynamic correlation effects.

In principle, dynamical exchange effects in the dielectric function are described by the time-dependent Hartree-Fock (TDHF) equation. However, the solution of this equation is far from obvious. In the present paper, we do not give a full derivation of the TDHF equation, but only summarize the underlying assumptions. (A detailed derivation in the notations used here can e.g. be found in [2]).

Suppose one applies an external potential, with Fourier components $\Phi_{\vec{q},\omega}$ and corresponding Fourier components $n_{\vec{q},\omega}$ of the induced electron density. The dielectric function is defined as the ratio between the applied potential and the total potential (which is the sum of the induced and the applied potential):

$$e\Phi_{\vec{q},\omega} + \frac{4\pi e^2}{q^2} n_{\vec{q},\omega} = \frac{e\Phi_{\vec{q},\omega}}{\epsilon(q,\omega)} \tag{B.2}$$

(In this equation, explicit use has been made of the translational symmetry of the electron gas; for inhomogeneous systems the dielectric **matrix** has to be studied).

To derive the induced density, a useful quantity is the Wigner distribution function $f_\sigma(\vec{p},\vec{q},\omega)$, which is a function of the spin σ and the momentum $\vec{p}$ of the electrons, and of the wave vector $\vec{q}$ and the frequency ω of the applied field. The Wigner distribution function is a quantum analogue of the classical Boltzmann distribution function [12], from which the induced density can be derived:

$$n_{\vec{q},\omega} = \sum_\sigma \int d^3p \, f_\sigma(\vec{p},\vec{q},\omega) \tag{B.3}$$

For a derivation of the equation of motion, including dynamical exchange effects, we refer to Ref. [2], where also the details are given on the Hartree-Fock decoupling in the equation of motion. (The interaction is supposed to be switched on adiabatically, which in the following equations is reflected in the frequency contributions with an infinitesimal positive constant: $\omega^+ = \omega + i\delta$). The resulting equation is:

$$(\omega^+ - \vec{p}.\vec{q}/m) f_\sigma(\vec{p},\vec{q},\omega) = -\frac{1}{2} N_{\vec{q}}(\vec{p}) U_{\vec{q},\omega} + X_\sigma(\vec{p},\vec{q},\omega) \tag{B.4}$$

where $U_{\vec{q},\omega}$ is the Hartree potential:

$$U_{\vec{q},\omega} = e\Phi_{\vec{q},\omega} + \frac{4\pi e^2}{q^2} n_{\vec{q},\omega} \tag{B.5}$$

and $X_\sigma(\vec{p},\vec{q},\omega)$ accounts for the exchange effects:

$$X_\sigma(\vec{p},\vec{q},\omega) = \int d^3p' \frac{4\pi e^2 \hbar^2}{|\vec{p} - \vec{p}\,'|^2} [N_{\vec{q}}(\vec{p}) f_\sigma(\vec{p}\,',\vec{q},\omega) - N_{\vec{q}}(\vec{p}\,') f_\sigma(\vec{p},\vec{q},\omega)] \tag{B.6}$$

In these equations, $N_{\vec{q}}(\vec{p})$ describes the difference of two equilibrium distributions, shifted over a wave vector $\vec{q}$ with respect to each other:

$$\hbar N_{\vec{q}}(\vec{p}) = f^0(\vec{p} + \frac{\hbar\vec{q}}{2}) - f^0(\vec{p} - \frac{\hbar\vec{q}}{2}) \tag{B.7}$$

where $f^0(\vec{p})$ is the Fermi-function properly normalized (nl. $f^0(\vec{p}) = 2/(2\pi\hbar)^3$ if $p < p_F$, and zero elsewhere).

The exchange term $X_\sigma(\vec{p},\vec{q},\omega)$ substantially complicates the task to solve the TDHF-equation (B.4). If this term is neglected, the solution is the Lindhard distribution function $f_\sigma^L(\vec{p},\vec{q},\omega)$:

$$f_\sigma^L(\vec{p},\vec{q},\omega) = -\frac{1}{2} \frac{e\Phi_{\vec{q},\omega}}{1 + Q_0(q,\omega)} \frac{N_{\vec{q}}(\vec{p})}{\omega^+ - \vec{p}.\vec{q}/m} \tag{B.8}$$

$$Q_0(q,\omega) = \frac{4\pi e^2}{q^2} \int d^3p \frac{N_{\vec{q}}(\vec{p})}{\omega^+ - \vec{p}.\vec{q}/m} \tag{B.9}$$

The resulting dielectric function is the RPA-dielectric function:

$$\epsilon^{RPA}(1,\omega) = 1 + Q_0(q,\omega) \qquad (B.10)$$

Although the integral in Eq.(B.10) can be evaluated analytically, we prefer to give its integral representation because of the apparent structure. The denominator accounts for the difference in energy between the initial and final state when a perturbation of frequency ω is applied, and in which the unperturbed states are assumed to be free-particle states.

The analytical solution of the TDHF equation (B.4) is beyond the scope of the known rigorous methods. But also the numerical solution involves several problems. Since the TDHF equation is a three-dimensional integral equation, the calculations tend to be rather time consuming, even if the kernel would be smooth. Taking into account that three free parameters enter the problem (nl. the wave vector $\vec{q}$ and the frequency ω of the perturbation, and the average electron density), this three-dimensional integral equation has to be solved for a quite extended range of parameter values.

Furthermore, the kernel is far from smooth, and includes simultaneously the rather singular Fourier transform of the Coulomb potential, as well as a resonance if the excitation frequency equals the difference in the single-particle energies. In order to make this combination of problems tractable, some transformations of the TDHF equation are required.

The introduction of dimensionless variables by using the appropriate Fermi units

$$\vec{k} = \frac{\vec{q}}{k_F} \quad ; \quad \nu = \frac{\hbar\omega}{2E_F} \quad ; \quad \xi = \frac{\vec{p}}{\hbar k_F} \qquad (B.11)$$

substantially simplifies the TDHF equation (B.4-B.7), and allows to collect all the density-dependent terms in a single parameter:

$$C \equiv 2\pi^2 \left(\frac{9\pi}{4}\right)^{1/3} \frac{1}{r_s} \qquad (B.12)$$

Furthermore, we explicitly perform the integration in the second term of (B.6), which accounts for the exchange contribution to the single-particle energy in the Hartree-Fock approximation, and combine the result with the factor $(\omega^+ - \vec{p}.\vec{q}/m)$ in the left hand side of Eq. (B.4). Although the two exchange terms are separated from each other in this way, this presents no problems if one aims at a rigorous solution; for approximate treatments, this separation might introduce violations of sum rules if both terms are not treated to the same order of approximation.

To simplify the notation, we also consider the equation of motion for $f_\sigma(\vec{p}+\frac{\hbar\vec{q}}{2},\vec{q},\omega)$ instead of $f_\sigma(\vec{p},\vec{q},\omega)$, and express the Hartree term $U_{\vec{q},\omega}$ in terms of the dielectric function via the induced density. Collecting terms after all these algebraic transformations, the TDHF equation becomes:

$$b_{\vec{k},\nu} = 1 - \frac{1}{C}\int_V d^3\xi'\left(\frac{1}{\left|\vec{\xi}+\vec{\xi}'+\vec{k}\right|^2}\frac{b^*_{\vec{k},-\nu}(\vec{\xi}')}{\left[(-\nu)+-\Lambda_{\vec{k}}(\vec{\xi}')\right]^*} + \frac{1}{\left|\vec{\xi}+\vec{\xi}'\right|^2}\frac{b_{\vec{k},\nu}(\vec{\xi}')}{\nu+-\Lambda_{\vec{k}}(\vec{\xi}')}\right) \tag{B.13}$$

where

$$\Lambda_{\vec{k}}(\vec{\xi}) = \frac{k^2}{2} + \vec{\xi}.\vec{k} - \frac{2\pi}{C}\left[T(|\vec{\xi}+\vec{k}|) - T(\xi)\right] \tag{B.14}$$

$$T(\xi) = \frac{1-\xi^2}{2\xi}\log\left|\frac{1+\xi}{1-\xi}\right| \tag{B.15}$$

and where $b_{\vec{k},\nu}(\vec{\xi})$ is proportional to the Wigner distribution function, but with its resonant behavior in the excitation frequency extracted:

$$b_{\vec{k},\nu}(\vec{\xi}) = \frac{4\pi e^2\hbar^2 k_F}{e\Phi_{\vec{q},\omega}}\epsilon(q,\omega)\left[\nu^+ - \Lambda_{\vec{k}}(\xi)\right] C f_\sigma(\vec{p}+\frac{\hbar\vec{q}}{2},\vec{q},\omega) \tag{B.16}$$

The integration volume V in (2.3) is determined by the condition:

$$\vec{\xi}\in V \quad\Longleftrightarrow\quad |\vec{\xi}|<1 \;\text{ and }\; |\vec{\xi}+\vec{k}|>1 \tag{B.17}$$

The dielectric function (B.2) can immediately be expressed in terms of the smooth function $b_{\vec{k},\nu}(\vec{\xi})$, by using (B.3) in combination with (B.16) and the dimensionless variables introduced in (B.11-B.12):

$$\epsilon(q,\omega) = 1 - \frac{2}{k^2}\frac{1}{C}\int d^3\xi'\left(\frac{\xi'_\perp b^*_{\vec{k},-\nu}(\vec{\xi}')}{\left[(-\nu)+-\Lambda_{\vec{k}}(\vec{\xi}')\right]^*} + \frac{\xi'_\perp b_{\vec{k},\nu}(\vec{\xi}')}{\nu+-\Lambda_{\vec{k}}(\vec{\xi}')}\right) \tag{B.18}$$

Since both (B.16) and (B.18) are axial-symmetric with respect to rotations around the wave vector $\vec{k}$ of the applied excitation, the azimuthal integration can be performed:

$$b_{\vec{k},\nu}(\vec{\xi}) = 1 - \frac{2\pi}{C}\int_S d^2\xi'\left(\frac{\xi'_\perp j_{\vec{\xi}}(-\vec{\xi}'-\vec{k})b^*_{\vec{k},-\nu}(\vec{\xi}')}{\left[(-\nu)+-\Lambda_{\vec{k}}(\vec{\xi}')\right]^*} + \frac{\xi'_\perp j_{\vec{\xi}}(\vec{\xi}')b_{\vec{k},\nu}(\vec{\xi}')}{\nu+-\Lambda_{\vec{k}}(\vec{\xi}')}\right) \tag{B.19a}$$

with

$$j_{\vec{\xi}}(\vec{\xi}') = \frac{1}{\sqrt{\left[(\xi_z-\xi'_z)^2+\xi_\perp^2+\xi'^2_\perp\right]^2 - 4\xi_\perp^2\xi'^2_\perp}} \tag{B.19b}$$

and

$$\epsilon(q,\omega) = 1 - \frac{2}{k^2}\frac{2\pi}{C}\int_S d^2\xi'\left(\frac{\xi'_\perp b^*_{\vec{k},-\nu}(\vec{\xi}')}{\left[(-\nu)+-\Lambda_{\vec{k}}(\vec{\xi}')\right]^*} + \frac{\xi'_\perp b_{\vec{k},\nu}(\vec{\xi}')}{\nu+-\Lambda_{\vec{k}}(\vec{\xi}')}\right) \tag{B.20}$$

One is thus left with an integral equation in two variables, where the two-dimensional integration domain S is defined by:

$$\vec{\xi} \in S \iff |\vec{\xi}| < 1 \text{ and } |\vec{\xi} + \vec{k}| > 1 \qquad (B.21)$$

In this integration domain, $\xi_\perp$ and ξ_z denote the components of $\vec{\xi}$ perpendicular resp. parallel to $\vec{k}$. The integration domain is thus a semicircle of unit radius centered at the origin, of which in the case $k < 2$ one has to eliminate the overlap region with a semicircle, centered around $\xi_z = -k$.

The integral equation (B.19) has to be solved numerically.*The numerical procedure used was discussed in section 2 of the present paper. For a discussion of the results, we refer to Ref. [1]. The main purpose of the present paper is to show the basic idea of the **global vectorisation**, realized in the numerical solution of the TDHF equation, of which a basic example was given detail. From the description of the procedure, it is clear that global vectorisation of the program is relatively easy along similar lines.

ACKNOWLEDGEMENT

Financial support from the Supercomputer Project of the National Fund for Scientific Research (Belgium) is gratefully acknowledged.

REFERENCES

[1] F. Brosens, J. T. Devreese, phys. stat. sol. (b) **147**, 173–183 (1988).

[2] F. Brosens, L. F. Lemmens and J. T. Devreese, phys. stat. sol. (b)**74**, 45 (1976); F. Brosens, J. T. Devreese and L. F. Lemmens, phys. stat. sol. (b)**81**, 99 (1977); J. T. Devreese, F. Brosens and L. F. Lemmens, Phys. Rev. **B21**, 1349 (1980); F.Brosens, J. T. Devreese and L. F. Lemmens, Phys. Rev. **B21**, 1366 (1980); F. Brosens and J. T. Devreese in "Proceedings of the 1981 NATO Advanced Study Institute on Electron Correlations in Solids, Molecules and Atoms", Eds. J. T. Devreese and F. Brosens (Plenum, New York, 1983), p. 143; F. Brosens and J. T. Devreese, Helv. Phys. Acta **56**, 223 (1983).

[3] C. H. Reinsch, Numer. Math. **10**, 177 (1967)

[4] J. Lindhard, Kong. Danske Vid. Selsk., Mat.-fys. Medd. **24**, nr. 8, (1954).

[5] D. Bohm and D. Pines, Phys. Rev. **92**, 609 (1953).

* Note that in the limit $r_s \to 0$, the RPA result is obtained. As is clear from (B.19a) and (B.12), this limit gives $b_{\vec{k},\nu}(\vec{\xi}) \to 1$. Since $\lim_{r_s \to 0} \Lambda_{\vec{k}}(\vec{\xi}) = k^2/2 - \vec{\xi}.\vec{k}$, (B.20) then indeed gives the RPA dielectric function. One thus might feel tempted to consider $b_{\vec{k},\nu}(\vec{\xi}) = 1$ as a zero-order value for an iterative procedure to solve (B.19a). Unfortunately, experience showed that successive approximations for $b_{\vec{k},\nu}(\vec{\xi})$ oscillate with increasing amplitude as the number of iterations increases.

[6] P. E. Batson, C. H. Chen and J. Silcox, Phys. Rev. Lett. **37**, 937 (1976).

[7] P. C. Gibbons, S. E. Schnatterly, J. J. Ritsko and J. R. Field, Phys. Rev. **B13**, 2451 (1976).

[8] S. Schnatterly in "Proceedings of the 1981 NATO Advanced Study Institute on Electron Correlations in Solids, Molecules and Atoms", Eds. J. T. Devreese and F. Brosens (Plenum, New York, 1983), p. 1;

[9] A. J. Glick and R. A. Ferrell, Ann. of Phys. **11**, 359 (1960).

[10] J. Hubbard, Proc. Roy. Soc. A**240**, 539 (1957); A**243**, 336 (1958).

[11] For an overview: see e.g. G. Niklasson in in "Proceedings of the 1981 NATO Advanced Study Institute on Electron Correlations in Solids, Molecules and Atoms", Eds. J. T. Devreese and F. Brosens (Plenum, New York, 1983), p. 99;

[12] W. E. Brittin and W. E. Chappell, Rev. Mod. Phys. **34**, 620 (1962).

[13] M. C. Dharma-wardana and R. Taylor, J. Phys. F: Metal Phys. **10**, 2217 (1980).

[14] A. W. Overhauser, Phys. Rev. **B2**, 874 (1970).

APPLICATION OF VECTORIZATION ON AB-INITIO CALCULATIONS OF SILICON CARBIDE AND BORON NITRIDE

P.E. Van Camp and J.T. Devreese*

University of Antwerp (RUCA), Groenenborgerlaan 171, B-2020 Antwerpen, Belgium

I. INTRODUCTION

It is only during the past decade that some electronic and structural properties of solids have been calculated successfully starting from first principles [1]. The underlying theory was developed during the mid sixties, but computers at that time were far too small and slow to handle this problem.

The principal aim of the ab-initio theory is the calculation of the observable quantities based solely on the fundamental laws of quantum mechanics. In the case of the electronic and structural properties of solids the only input consists of the atomic numbers of the atoms from which the crystal is built. Of course also the fundamental physical constants (the electronic charge and mass, Planck's constant, etc.) are needed. Since one cannot treat the basically infinite number of crystal lattices one is limited to investigate a small number of them, determined on general physical and chemical grounds. It should be stressed that nowhere in the calculation parameters are fitted to experimental values.

In this article the ab-initio non-local pseudopotential method is applied to the semiconducting materials silicon carbide and boron nitride. Both compounds are treated in their most stable form, i.e. in the zincblende structure. The calculated quantities are:

1. The equilibrium lattice constant. The total energy of the crystal is determined for several volumes (i.e. pressures). The zero pressure condition then gives the calculated equilibrium lattice constant.
2. The bulk modulus and its pressure derivative are also determined from the total energy versus volume relation (i.e. the equation of state).
3. The electronic charge density.

---

* Also at: Department of Physics, University of Antwerp (UIA), Universiteitsplein 1, B-2610 Wilrijk-Antwerpen, Belgium and University of Technology, P.O. Box 513, NL-5600 MB EINDHOVEN, The Netherlands.

4. The one-particle band-energies and their first and second order pressure derivatives.

All these quantities can in principle be measured and some of them have been indeed been measured for the compounds under investigation.

II. THEORETICAL FRAMEWORK

In the local density approximation [2] the total crystalline energy is given by:

$$E_T[\rho] = \sum_{val} E_{kn} - \frac{1}{2}\int d\vec{r} \int d\vec{r}\,' \frac{\rho(\vec{r})\rho(\vec{r}\,')}{|\vec{r}-\vec{r}\,'|}$$

$$+ \int dr\, \rho(\vec{r})\, [\varepsilon_{xc}(\rho)-\mu_{xc}(\rho)] + \sum_{\substack{la\\mb}} \frac{Z_a Z_b}{|\vec{R}_{la}-\vec{R}_{mb}|} \quad (1)$$

The first part contains the valence band energies E_{kn}, the second part is the Hartree term, the third the exchange correlation contribution and the last one is the pure electrostatic energy of the ions. The one-particle energies E_{kn} and the electronic charge density $\rho(\vec{r})$ are obtained from the Kohn-Sham equations [3] (in atomic units):

$$[-\frac{1}{2}\nabla^2 + V_i(\vec{r}) + \int d\vec{r}\,' \frac{\rho(\vec{r}\,')}{|\vec{r}-\vec{r}\,'|} + \mu_{xc}(\rho)]\psi_{kn} = E_{kn}\psi_{kn} \quad (2)$$

The ionic potential $V_i(\vec{r})$ is taken to be the ab-initio nonlocal pseudopotential as tabulated e.g. in ref. [4]. For the exchange-correlation potential μ_{xc} the Wigner interpolation formula [5] is used:

$$\mu_{xc}(\rho) = \frac{d}{d\rho}(\rho\, \varepsilon_{xc}(\rho)) \quad (3)$$

and

$$\varepsilon_{xc}(\rho) = -\frac{3}{4}\left(\frac{3}{\pi}\right)^{1/3} \rho^{1/3} - \frac{0.056\, \rho^{1/3}}{0.079+\rho^{1/3}} \quad (4)$$

In the present work the Kohn-Sham equations (Eq. (2)) are solved using a plane wave basis [6], i.e.

$$\psi_{kn}(\vec{r}) = \sum_{G} C_{kn}(\vec{G})\, e^{i(\vec{k}+\vec{G})\vec{r}} \quad (5)$$

with $\vec{G}$ a reciprocal lattice vector. The system of coupled integro-differential equations is thereby converted into an eigenvalue problem

$$HC = EC \quad (6)$$

that has to be solved iteratively since H depends upon the solution C.

The equation of state of the system

$$P = -\frac{dE_T}{dV} \quad (7)$$

can then be determined using the calculated total energies (Eq. (1)) for different volumes (i.e. different values of the lattice constant). In order to avoid a numerical differentiation the Murnaghan equation of state [7]

$$P = \frac{B_o}{B_o'} \left[\left(\frac{V_o}{V}\right)^{B_o'} - 1 \right] \qquad (8)$$

is fitted to the calculated points. From this fit the equilibrium volume V_o (i.e. the lattice constant), the bulk modulus B_o and its pressure derivative B_o' are obtained. It should be noted that other functional forms of the equation of state give nearly identical results for a, B_o and B_o'.

III. COMPUTATIONAL DETAILS

A. Input Parameters

In this paper the above summarized theory is applied to SiC and BN. Previously, the present authors have treated, in the same theoretical framework, the covalent semiconductors [8] and the Ga-compounds [9]. A more complete account of the SiC and BN calculation was published before [10]. Both compounds crystallize in the zincblende (T_d^2) structure. Other modifications exist at higher pressure and temperature. The parameter controlling the number of plane waves in the expansion Eq. (5), the so-called kinetic energy cut-off, used is 37 Ry for SiC and 55 Ry for BN. This amounts to respectively 531 and 537 plane waves. The summation over all valence states in the Brillouin zone (i.e. the first part of Eq. (1)), is performed employing the "special points scheme". Tests using 2 or 10 points have shown that there is no need to take a finer mesh.

The total energy is determined for 8 different volume and least squares fitted to the Birch equation of state.

B. Numerical Procedures

There are mainly three numerical procedures involved in the present calculation:

- the computation of the exchange-correlation potential (μ_{xc} in Eq. (2)) by means of Fast Fourier Transform techniques
- the iteration scheme needed to solve the Kohn-Sham equations (or rather their matrix equivalent) self-consistently
- the diagonalization of a large matrix

Time-wise the matrix diagonalization is important as is the iteration procedure. The time needed to do the Fast Fourier Transform on the other hand can be neglected.

1. The Fast Fourier Transform. The Fast Fourier Transform (FFT) algorithm [11] is well-known and extensively used in a variety of problems. Due to the fact that the charge density is complex, we use in our calculation an N-point complex to complex FFT. There exist several standard library implementations for serial computers, one of which is used in this work. It should be noted that recently the algorithm was vectorized

[12], resulting in a speedup of 30% over an assembly-coded library subroutine.

2. The self-consistent iteration scheme. The total run time depends linearly on the number of iterations needed to determine the charge density. As such it is very important to find an efficient iteration scheme. In the literature [13-15], there is a variety of schemes available, one of the simplest and most frequently used is the "simple mixing scheme" [14].

The equation to be solved can then be written formally as

$$\rho_n = F[\rho_{n-1}] \tag{9}$$

where the index n labels the iteration count and F is a complicated nonlinear operator. Unfortunately, as it stands, the simple iteration Eq. (9) often starts giving oscillating charge densities and finally diverges. So convergence is not guaranteed and is usually slow. The oscillation produced by Eq. (9) can be damped by modifying the equation to

$$\rho_n = \alpha\, F[\rho_{n-1}] + (1-\alpha)\, \rho_{n-1} \tag{10}$$

where α is a constant mixing parameter ($\neq 0$, but otherwise undetermined). Experience has shown that in order to obtain convergence one has to use small values for α resulting in the need for many iterations. This scheme has been modified to determine an optimal mixing parameter in each iteration step [15]. This improved iteration scheme, which is used in the present work, has several advantages over the simple mixing scheme:

- there is no need to guess α, it is calculated
- the present authors have, up to now, not encountered a case where the scheme diverges, contrary to the simple mixing scheme
- the convergence is faster than for the simple mixing scheme. A possible exception occurs when the starting value of ρ is close to the converged value.

3. Matrix diagonalization. Central to the whole calculation is the diagonalization of the Hamiltonian matrix, which in the present case is complex hermitian. To determine the total energy of the crystal, one only needs the valence band energies and wavefunctions. Mathematically speaking we want a small number (in the present case 4) of the lowest eigenvalues and corresponding eigenvectors of a large hermitian matrix (here with dimension up to 537). Work in progress on related materials indicates that e.g. for the other nitrides a larger basis (up to 1000 plane waves) is needed. Therefore an efficient algorithm to find a small number of the lowest eigenvalues and eigenvectors should be used.

One of the best algorithms available to achieve this uses the following strategy [16]:

- reduction of the given matrix to real symmetric tridiagonal form (i.e. the Householder transformations)
- some of the eigenvalues of the tridiagonal matrix are obtained using the bisection method. This method is more efficient than the implicit QL-method since only a small number of eigenvalues is wanted

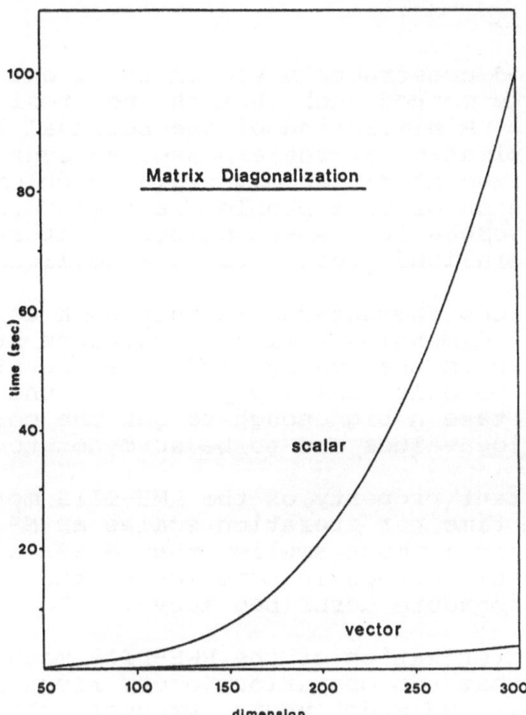

Fig. 1. Time (in sec.) for matrix diagonalization in scalar and vector mode as a function of the dimension.

- the corresponding eigenvectors of the tridiagonal matrix are calculated using inverse iteration
- these eigenvalues are then backtransformed to yield the eigenvectors of the original matrix.

This algorithm cannot be vectorized completely. The reduction is partly vectorized as is the inverse iteration. In Figure 1 we show timings for this algorithm on a CYBER-205 (2-pipe) in scalar and in vector mode for a real test matrix. From the figure it is obvious that the scalar time rises much more rapidly than the vector time. The ratio scalar/vector for a dimension of 300 is already 25.5.

For dimensions above ± 200 this algorithm can be replaced by a more efficient one: the "residual minimization/direct inversion in iterative subspace" method (RMS-DIIS) [17]. This algorithm is well suited to find the lowest in eigenvalues and eigenvectors of a NxN hermitian matrix. Typically n is a small fraction of N here (< 2%). There are several iterative methods available in the literature [17-19]. In principle they all start from a guessed eigenvector and generate by some procedure a new eigenvector. Usually the starting vector is obtained from the original matrix truncated to a much smaller dimension. Care should be taken that the small matrix gives the correct ordering and degeneracies of the wanted eigenvalues. The main difference between the iterative methods is the strategy by which a new guess vector is constructed. The measure of the extent to which an approximate eigenvector x_i and corresponding eigenvector λ_i fail to be exact is the residual vector R:

$$R = (H-\lambda_i)|x_i> \qquad (11)$$

The RMS-DIIS method constructs a vector increment δx_i using a Newton-Raphson like method such that the new residual is as small as possible. Minimization of the residual leads to a new eigenvector approximation at the expense, however, of the solution of a generalized hermitian eigenvalue problem. Fortunately, the dimension of this problem is the iteration number. If the method converges in a small number of iterations the solution of the generalized problem takes a negligible time.

For the matrices encountered in this work using a carefully chosen small dimension n (i.e. a complete star) convergence is reached in on the average 10 iteration steps. This number appears to be only weakly dependent on the size of n. However, one must take n big enough to get the correct ordering of the lowest n eigenvalues and to be sure not to miss any.

A very important property of the RMS-DIIS method is that the computational time per iteration scales as N^2. Since the number of iterations is much smaller than N and almost constant it is clear that this method is superior to the standard matrix diagonalization procedure described above.

As far as vectorization of the RMS-DIIS method is concerned one notes that two operations occur frequently: a vector-matrix multiply and a dot vector product. It is clear that they can both be vectorized easily. As an example Figure 2 shows timings of a full matrix-matrix multiplication on a

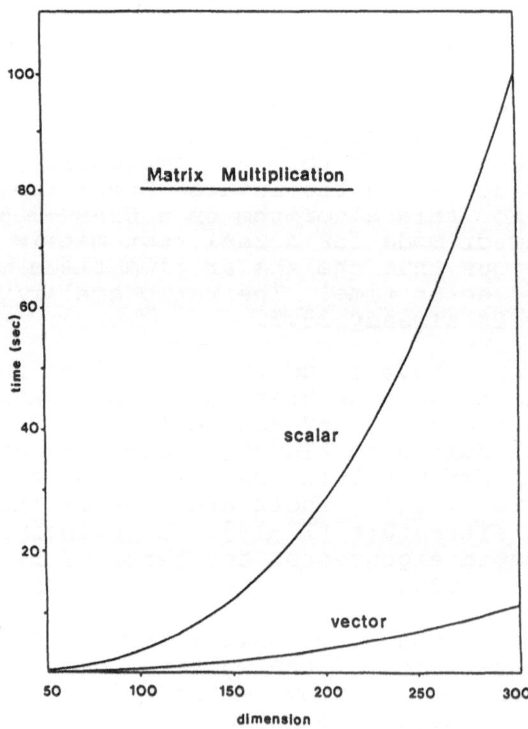

Fig. 2. Time (in sec.) for matrix multiplication in scalar and vector mode as a function of the dimension.

Table I. Lattice constant, bulk modulus and its pressure derivative for SiC and BN compared with experimental values.

	SiC		BN	
	Theory	Experiment	Theory	Experiment
a (Å)	4.3925	4.36 [20]	3.6492	3.615 [20]
B_o (Mbar)	1.9621	2.24 [21]	3.6662	4.65 [22]
B_o'	3.5722	-	2.9113	-

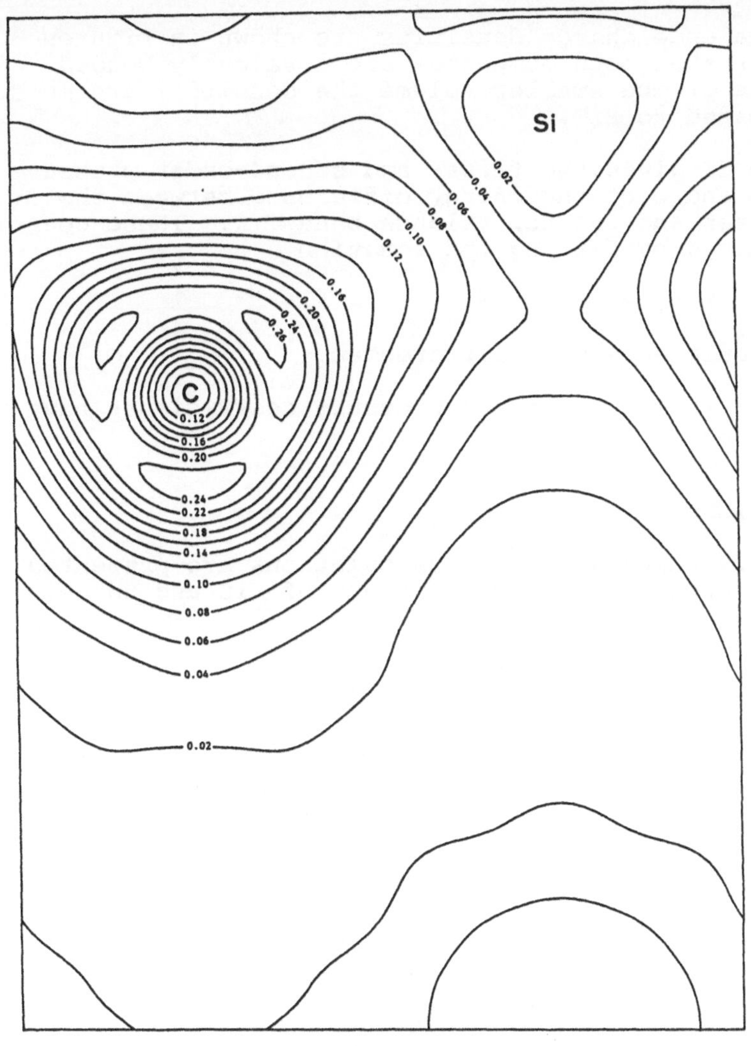

Fig. 3. Valence charge density (in e/au) in the (110) plane of SiC. The contour step is 0.02 e/au.

CYBER-205 in scalar and in vector mode. In this case a real test matrix has been used. In the scalar code the loops were unrolled to a depth of 4. There is, however, one point that partly inhibits vectorization of the vector-matrix multiply. Because of the hermiticity of the Hamiltonian matrix only the lower triangular part is stored. As is often the case a vectorized algorithm requires much more memory than a scalar one. The difficult part is to find the balance between using huge amounts of memory (resulting in expensive page faults) and only be able to partly vectorize the algorithm. In our programs the latter way has been chosen.

IV. RESULTS

The Murnaghan equation of state was fitted to the total energy calculated at eight different lattice constants. The resulting values of a, B_o and B_o' are given in Table I, together with the available experimental information.

The valence charge densities are shown in Figures 3 and 4. It is clear that both compounds are covalently bonded. As a consequence of the smaller volume the densities are much higher in BN compared to SiC.

Table II gives the first- and second-order pressure coefficients b and c of the energy difference between the Γ_1, X_1 and L_1 states and the top valence band Γ_{15}. These coefficients are determined by fitting the equation

$$E(p) - E(o) = bp + cp^2 \tag{12}$$

to the calculated energy differences.

To our knowledge, these pressure coefficients have not yet been measured.

V. CONCLUSIONS

In this paper ab-initio calculations are presented of electronic properties under pressure of SiC and BN using norm

Table II. Pressure coefficients of the direct and indirect X and L band gaps (b in eV/Mbar and c in eV/Mbar^2).

	SiC		BN	
	b	c	b	c
Γ	6.17	-3.62	0.86	-1.86
X	-0.11	3.37	0.26	-0.08
L	4.76	-1.21	3.12	-2.69

conserving non-local pseudopotentials in the local density approximation. Besides the lattice constant, the bulk modulus and its pressure derivative, band energies and first and second order pressure coefficients are determined. These calculations

Fig. 4. Valence charge density (in e/au) in the (110) plane of BN. The contour step is 0.02 e/au.

are time consuming and a lot of effort has been invested in vectorizing the programs and in adapting the implementations to run on a pipelined computer. The matrix diagonalization procedures specially adapted to treat large matrices are described in this paper.

ACKNOWLEDGMENTS

The authors would like to thank Dr. V.E. Van Doren, who is also involved in this research.

This work was performed in the framework of the "Institute for Materials Science" (I.M.S.) of the University of Antwerp (RUCA and UIA) funded by the IUAP 11 (Interuniversitaire Attractiepool 11 'Materials Science') of the Belgian Ministry of Scientific Affairs. CYBER-205 supercomputer time was provided by the "NFWO - Supercomputer Project" of the Belgian National Science Foundation (NFWO).

REFERENCES

1. J.T. Devreese and P.E. Van Camp, eds., "Electronic Structure, Dynamics and Quantum Structural Properties of Condensed Matter", Plenum Press, New York (1985), and references therein.
2. P. Hohenberg and W. Kohn, Phys. Rev. 136:3864 (1964).
3. W. Kohn and L.J. Sham, Phys. Rev. 140:A1133 (1965).
4. G.B. Bachelet, D.R. Hamann and M. Schlüter, Phys. Rev. B26:4199 (1982).
5. E. Wigner, Phys. Rev. 46:1002 (1934).
6. M.T. Yin and M.L. Cohen, Phys. Rev. B26:5668 (1982).
7. F.D. Murnaghan, Proc. Natl. Acad. Sci. USA 3:244 (1944).
8. P.E. Van Camp, V.E. Van Doren and J.T. Devreese, Phys. Rev. B34:1314 (1986); ibid. Physica Scripta 35:706 51987); ibid. Phys. Rev. B38 (December 15, 1988).
9. P.E. Van Camp, V.E. Van Doren and J.T. Devreese, Phys. Rev. B38:9906 (1988).
10. P.E. Van Camp, V.E. Van Doren and J.T. Devreese, Phys. Stat. Sol. (b) 146:573 (1988).
11. J.W. Cooley and J. Tukey, Math. Comp. 19:297 (1965).
12. D.H. Bailey, J. Supercomp. 1:43 (1987); ibid. Int. J. Supercomp. Appl. 2:82 (1988).
13. M.S. Dewar and P.K. Weiner, Comp. Chem. 2:31 (1978).
14. P. Dederichs and R. Zeller, Phys. Rev. 28:5462 (1983).
15. R. Badziag and F. Solms, Comp. Chem. 12:233 (988).
16. B. Garbow, J. Boyle, J. Dongarra and C. Moler, "Lecture Notes in Computer Science", Volume 51 (1977).
17. P. Bendt and A. Zunger, Bull. Am. Phys. Soc. 27:248 (1982); D.M. Wood and A. Zunger, J. Phys. A18:1343 1985); D.L. Martins, private communication.
18. B.N. Parlett, "The Symmetric Eigenvalue Problem", Prentice Hall (1980).
19. E.R. Davidson, J. Comp. Phys. 17:87 (1975).
20. Landolt-Börnstein, "Numerical Data Group III", Volume 22a, Springer Verlag, Heidelberg (1987).
21. D. Yean and J. Riter, J. Phys. Chem. Sol. 32:653 (1971).
22. J. Sanjurgo, E. Lopez-Cruz, P. Vogl and M. Cardona, Phys. Rev. B28:4579 (1983).

APPLICATIONS OF SUPERCOMPUTERS IN MATHEMATICS

Herman J.J. te Riele

Centre for Mathematics and Computer Science
Kruislaan 413, NL-1098 SJ Amsterdam, The Netherlands

ABSTRACT

This paper provides a concise survey of the role played by vector and parallel computers in the solution of problems in computational mathematics. Some vectorization and parallelization techniques are discussed. Many examples illuminate the discussion.

1. INTRODUCTION

The advent of extremely powerful computers like the CRAY-1 in 1976 and the CYBER 205 in 1981 has strongly stimulated the interest of scientists and engineers in finding ways to (re-) organize their algorithms such that these computers can solve their problems with maximal performance. One could say that pipelining and parallelism in hardware has added a new dimension to algorithm design and analysis.

The CRAY-1, CYBER 205, FACOM VP-200, HITACH S810 are examples of so-called pipelined or vector processors, which perform in an optimal way when they operate on long vectors of data. Reformulation of algorithms in terms of (long) vectors is called vectorization. In the past few years parallel computers, i.e., computers with more than one central processor which can operate independently and concurrently, have appeared (like the CRAY X-MP and the Denelcor HEP). For these machines it is important to parallelize algorithms, i.e., to trace (possibly different) subprocesses which can be executed independently.

Vectorization and parallelization are techniques which, of course, have much in common (cf. section 2.3). Therefore, we will only distinguish between the two concepts (and between vector and parallel computers) if this is necessary in the given context.

In the above development, one may perceive two trends: one is a tendency to optimize algorithms for a particular vector of parallel computer by exploiting specific hardware and software features of the machine; the other is to adapt and implement algorithms in such a way that the resulting software is portable and can be auto-vectorized by a good compiler. At this moment it is difficult to judge which approach is to be preferred: rapid developments in parallel hardware and in its price-performance ratio, and lack of standards in programming tools for vector and parallel computers are factors which make a definite choice difficult, if not impossible. Maybe the best choice at this moment is to "divide-and-conquer": develop portable software and if the performance obtained with auto-vectorization on a given machine is unsatisfactory, try to optimize the software for the given machine.

The solution of many problems in mathematics and physical sciences requires heavy computations. The corresponding algorithms can often be formulated in terms of operations on vectors and in terms of a (small or large) number of independent sub-computations. In particular, algorithms from numerical linear algebra, which operate on vectors and matrices, play a crucial role in many computational problems. A bibliography from Bochum (F.R.G.) (Bernutat-Buchmann et al., 1983) entitled: "Parallel Computing" illustrates the rapid developments: the total number of references in the second edition of September 1983 is 5161, against 2610 in the first edition of June 1982. Table 1 gives the "top ten" list of subjects from this bibliography. Here, hardware subjects show the highest scores (architecture, multiprocessor, vector computer). This is reflected in the kind of subjects covered by a number of recent conferences on supercomputers and applications (Duff and Reid, 1985; Emmen, 1985). Table 2 presents a list of main subjects from mathematics and physical sciences with their scores in the Bochum Bibliography.

In Section 2 of this paper, some general concepts concerning supercomputers and parallelism will be treated. Section 3 discusses a number of applications in mathematics and Section 4 treats important vectorization and parallelization techniques employed in these applications.

Table 1. The top ten subjects from Bernutat-Buchmann et al. (1983).

Subject	Number of references
Computer Architecture	1779
Algorithms	1177
Numerical Algorithms	680
Multiprocessors	592
Vector Computer	515
Networks	390
Image Processing	338
Complexity	324
Linear Algebra	303
Programming Languages	238

Table 2. Mathematical and Physical Sciences subjects from Bernutat-Buchmann et al. (1983).

Mathematics	#	Physical Sciences	#
Matrix Algorithms	192	Fluid Dynamics	69
PDEs	116	Pattern Recognition	68
FFT	111	Transonic Flow	35
Graph Theory	87	Air Traffic Control	33
Arithmetic Expressions	82	Potential Equation	31
Iterative Methods	73	Radar Control, Systems, Data Processing	24
Sorting	60		
Optimization	57	Poisson Equation	22
Matrix Multiplication	53	Weather Forecast	20
Sparse Matrices	47	Monte Carlo Method	18
Tridiagonal Matrices	41	Ballistic Missile Defense	14
ODEs	22	Navier-Stokes Equation	8
Direct Methods	17	Nuclear Physics	8
Runge Kutta Methods	4	Quantum Chemistry	5

Excellent surveys on vector and parallel computers and algorithms are: Heller (1978), Hockney and Jesshope (1981), Miranker (1971), Ortega and Voigt (1985), Sameh (1977), Schendel (1984), van Leeuwen (1983) and Zakharov (1984).

2. SOME GENERAL CONCEPTS CONCERNING VECTOR AND PARALLEL COMPUTING

2.1. Some Definitions

The speed of vector and parallel computers is often expressed in MFLOPS: the number of Million FLOating Point operations per Second. If a vector or parallel computer has a clock cycle time of c nanoseconds (e.g. c = 12.5 for the CRAY-1 and c = 20 for the CYBER 205) and if one result per cycle is produced (which is usually the case for the operations +, - and *), then the speed is 1000/c MFLOPS. However, these performances are difficult to reach in practice since there is always some overhead which decreases these figures. When we compare MFLOPS-speeds of different computers, we should be aware that different computers usually have different clock cycle times. For example, when two computers with different cycle times show the same MFLOPS-speed for some problem, then apparently the computer with the smallest clock cycle time shows the largest overhead.

When for a given problem we compare the CPU-times of a serial and a parallel or vector computer, then the speed-up S is defined as the quotient Ts/Tp where Ts is the serial and Tp the parallel CPU-time. According to Stone (1973), for a parallel processor with p processors, typical speed-up ratios are the following:

S	Examples of algorithms with this speed-up
kp	Matrix computations, mesh calculations
kp/log(p)	Sorting, tridiagonal linear systems, linear recurrence relations, polynomial evaluation
klog(p)	searching
k (independent of p)	Certain nonlinear recurrence relations, certain compiler processes

Here, k is a machine-dependent constant, independent of p. If S_p is the speed-up for p processors, then the efficiency E_p is defined as the quotient S_p/p. E_p measures how busy the parallel processors are during the computation. The longer the processors are idle, or carry out extra calculations introduced through the parallelisation of the algorithm, the smaller E_p becomes.

On the various architectures, the arithmetic operations may be executed in three different modes, viz. serial, pipelined and parallel. Consider, for example, the problem of adding two floating-point vectors $x = (x_i)$ and $y = (y_i)$, to obtain the sum vector $z = (z_i)$ $(i = 1,2,\ldots,n)$, where $z_i = x_i + y_i$. The operation of adding a pair x_i, y_i may be divided into four sub-operations, viz.,

(1) compare the exponents,
(2) shift,
(3) add mantissae, and
(4) normalize.

Figure 1 exemplifies the three different modes (derived from Hockney and Jesshope, 1981).

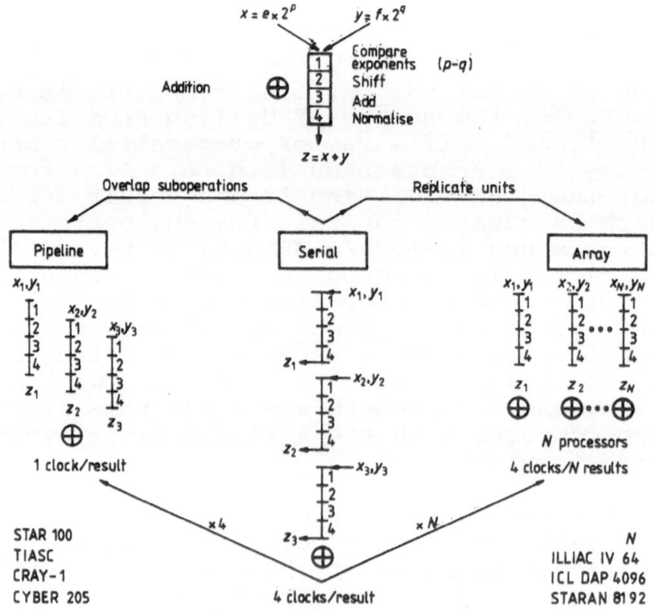

Fig. 1. Comparison of serial, pipelined and array architectures.

It may be interesting to remark that many supercomputers (like the CYBER 205) have both a vector and a scalar processor which may operate concurrently on different data. We have not yet seen any applications which exploit this feature of supercomputers.

2.2. Classification

Attempts have been made to arrange the various computer designs in classes. The simplest scheme is due to Flynn (1966): single (S) and multiple (M) streams of instructions (I) and data (D) are distinguished. This gives four possibilities: SISD, SIMD, MISD and MIMD. SISD is the classical von Neumann model: a single instruction stream operates on a single stream of data. SIMD is the class to which array processors and pipelined computers belong: all the processors interpret the same instructions and execute them on different data. The MISD class may be argued to be empty (cf. Schendel, 1984, p. 121). The MIMD class is the multi-processor version of the SIMD class: all processors interpret different instructions and operate on different data. For the four classes, Table 3 gives examples of machines, and schematically, examples of operations which can be executed at the same time.

A problem in this classification scheme are the pipelined processors. Usually, they are placed in the SIMD class although, strictly spoken, the instructions on different data are not executed at the same time; rather, each clock cycle one result is delivered from the input data stream(s).

A classification of parallel computers based on how computations proceed and how components in the architecture interact, is given in Table 4 (derived from Böhm, 1983). Other taxonomies of computers are given by Shore (1973) and by Schwartz (1983).

2.3. Algorithm Parallelism

It is customary (Hockney and Jesshope, 1981) to define, at any stage of an algorithm, the degree of parallelism of that algorithm as the number of independent operations that can be

Table 3. Flynn's classification.

type	operations which can be executed at the same time	examples
SISD	a + b	conventional von Neumann
SIMD	a + b, c + d	processor array (ICL/DAP, ILLIAC IV) pipelined processor (CRAY-1, CYBER 205)
MISD	a + b, a * b	
MIMD	a + b, c * d	multi-processors (CRAY X-MP, HEP)

Table 4. Computer architectures and their underlying computational model.

Model of computation	Corresponding Computer Architecture
A. Sequential control on scalar data.	A1. von Neumann-type computers A2. Multifunction CPU A3. Pipelined computers
B. Sequential control on vector data.	B1. Vector computers B2. Array processors
C. Independent, communicating processes	C1. Shared memory multiprocessors C2. Ultra computers C3. Networks of small machines
D. Functional and data-driven computation	D1. Reduction machines D2. Dataflow machines

performed in parallel, that is to say concurrently or simultaneously. On a pipelined computer the data would be interpreted as vectors and the operation would be performed on one vector. The parallelism is then the same as the vector length. On a processor array the data for each operation are allocated to different processing elements of the array and the operations on all elements are performed at the same time. The parallelism is then the number of data elements being operated upon in parallel in this way. The degree of parallelism may remain constant during the different steps of the algorithm, or it may vary from step to step.

Usually, there are two ways to analyse algorithms for use on vector and parallel processors:

1. Try to find, in a given algorithm, as many as possible independent subprocesses;

2. Devise a new algorithm with as many as possible independent subprocesses.

The following scheme suggests which type of computer is suitable, depending on whether the algorithm can be divided into few/many equal/different subprocesses.

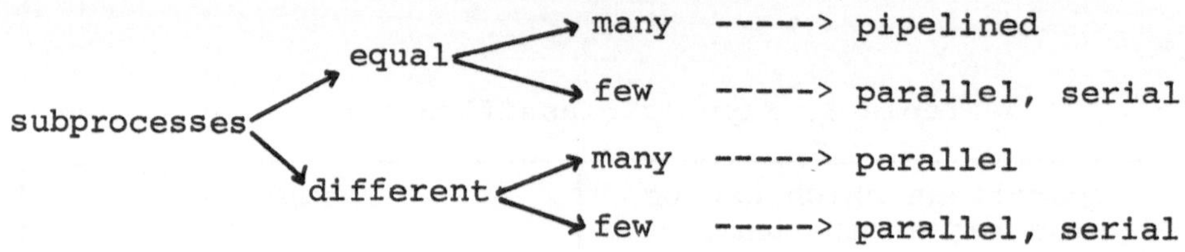

Various techniques for vectorization and parallelization are known, like recursive doubling, cyclic reduction, divide-and-conquer, pipelining and broadcasting. In fact, there is some overlap in these techniques. In Section 4 we shall explain the two most important ones, viz., recursive doubling and cyclic reduction. An interesting survey of many techniques, aimed at a theoretical analysis of parallel algorithms, was presented recently by Van Leeuwen (1983).

2.4. Organization of Data

In algorithms for parallel processing, the organization and dynamic arrangement of the data play a decisive role. Let us consider a very simple example of an SIMD processor with three processors P_1, P_2 and P_3, each of which has access to three storage locations. Suppose that the elements of a 3×3 matrix $A = (a_{ij})$ are stored in their "natural" order, as shown below:

P_1	P_2	P_3
a_{11}	a_{12}	a_{13}
a_{21}	a_{22}	a_{23}
a_{31}	a_{32}	a_{33}

Here, we assume that P_1 has access to a_{11}, a_{21} and a_{31} and P_2 and P_3 to the second and third columns of A, respectively. However, P_1 does not have access to the second and third column, and so on. Then, parallel operation is possible on the rows and main diagonals of A, but not on the columns of A. However, the following skew arrangement enables us to operate also on the columns:

P_1	P_2	P_3
a_{11}	a_{12}	a_{13}
a_{23}	a_{21}	a_{22}
a_{32}	a_{33}	a_{31}

Some general results concerning conflict free storage access in array processors are given by Schendel (1984).

A related, notorious problem, called memory bank conflict, may rise because of the presence of a so-called memory bank cycle time, which means that when loading an element from the memory bank, it is not possible to load another element from that same bank in the next few, e.g., three, clock cycles. For example, suppose we have an 8-bank machine and a vector is stored in the memory as follows: the elements with index $8m + n$, $0 \leq n \leq 7$, are stored in bank number n. Then, if we need the elements with indices 0, 1, 2, ... there will be no memory bank conflict and the speed of loading is one vector element per cycle. However, if we would need the elements with indices 0, 4, 8 ... there is a memory bank conflict and the speed of loading will be one element per two cycles. If we need the elements with indices 0, 8, 16, ... the loading speed will be only one element per three cycles. A remedy against memory bank conflict would be to store the elements in some skewed order. Of course, the best way to do this depends very much on the particular problem at hand.

2.5. Numerical Stability

Not much is known yet about stability, rounding errors and error propagation in parallel algorithms. In some cases, it appears that parallel processing leads to numerically inferior results, but this is not always the case. The following example shows how a parallel version of a simple algorithm actually yields better stability results than the serial version.

Consider the sum $S_N := \sum_{k=1}^{N} a_k$, where, for simplicity, we take $N = 2^n$. The serial algorithm for finding S_N reads as follows:

$$S_0 := 0,\ S_k := S_{k-1} + a_k\ ,\ k = 1,2,\dots,N.$$

If the mantissa of the floating point numbers has s binary places, then the machine approximation S_N' of S_N satisfies the inequality:

$$|S_N - S_N'| < 2^{-s}\ aN(N+1),$$

where $a = \max_k |a_k|$.

A parallel version of this algorithm reads as follows:

$$S_{0i} := a_i,\ i = 1,2,\dots,N$$

$$S_{ki} := S_{k-1,2i-1} + S_{k-1,2i}\ ,\ k = 1,2,\dots,n;\ i = 1,2,\dots,2^{n-k}$$

$$S_N := S_{n1}\ .$$

Here, estimation of the overall error yields:

$$|S_N - S_N'| < 2^{-s+1}\ aN\log_2 N\ ,$$

which improves the serial $0(N^2)$-upperbound to 0(NlogN). For this parallel algorithm, it is not difficult to compute, for a given number p of parallel processors, the speed-up S_p and efficiency E_p. For N = 8 and p = 2, 3 and 4, the results are given below.

p	S_p	E_p
4	7/3	7/12
3	7/4	7/12
2	7/4	7/8

3. APPLICATIONS

In the Bochum Bibliography (Bernutat-Buchmann et al., 1983), many fields of mathematics are mentioned in connection with parallel computing (cf. Table 2). Here, we shall discuss a number of important examples.

3.1. Solution of Systems of Linear Equations

An excellent survey of parallel linear algebra algorithms and their complexity is given by Heller (1978). He treats the following subjects:

* linear systems
  - general dense matrices
  - triangular systems
  - tridiagonal systems
  - block tridiagonal and band systems
  - sparse matrices

* eigenvalues

Presently, much research is carried out on vector and parallel algorithms in numerical linear algebra. We mention a few groups:

* Van der Vorst (Delft, The Netherlands)
* Dekker, Hoffmann (Amsterdam, The Netherlands)
* Axelsson (Nijmegen, The Netherlands)
* Evans (Loughborough, UK)
* Young (Houston, Texas, USA)
* Dongarra (Argonne, Illinois, USA)
* Sameh (Urbana, Illinois, USA)

Here, we shall briefly describe an algorithm for solving linear dense systems of equations on a CYBER 205, as presented by Hoffmann (1985). First some notational conventions: lower case Greek letters denote real scalars, lower case roman letters denote vectors and upper case letters stand for matrices. The j-th column of the matrix A is given as $a_{.j}$ and the i-th row as $a_{i.}$. The non-zero part of a column or row of a triangular matrix is indicated by writing a bar above the character which denotes the column or row. The order of a matrix is denoted by n. The algorithm used is the well-known Gaussian elimination process which is equivalent to the factorization of the coefficient matrix A into A = LDU (apart from pivoting). Here, U is an upper triangular, L a lower triangular and D a diagonal matrix, whose elements are denoted by ν_{ij}, λ_{ij} and δ_i, respectively. The elements α_{ij} of A satisfy the equations:

$$\alpha_{ij} = \sum_{k=1}^{\min(i,j)} \lambda_{ik} \delta_k \nu_{kj} \; ; \qquad i,j \in \{1,2,\dots,n\}.$$

In the following algorithm the matrices L and U (and the information for D) are built up in the location of A which should be clear from the notation. The choice of the diagonal elements of L, D and U is left open yet.

<u>for</u> k = 1(1)n <u>do</u>
<u>begin</u> determine $q \in \{k,\dots,n\} : |\alpha_{kq}| = \max_{k \le j \le n} |\alpha_{kj}|$

{search for maximum}

$a_{.q} \Leftrightarrow a_{.k}$ {interchange two length n columns}

$\alpha_{kk} \leftarrow \begin{Bmatrix} \nu_{kk} \\ \delta_k \\ \lambda_{kk} \end{Bmatrix} = ?$ {choose diagonal normalization, store}

$\bar{a}_{.k} \leftarrow \bar{\ell}_{.k} = \bar{a}_{.k}/(\delta_k \nu_{kk})$

{calculate k-th col of $\bar{L}$ and store}

for j = k+1(1)n do
begin $\alpha_{kj} \leftarrow \nu_{kj} = \alpha_{kj}/(\delta_k \lambda_{kk})$

{next element in $u_{k.}$ and store}

$a_{.j} \leftarrow a_{.j} - (\nu_{kj}\delta_k)\bar{\ell}_{.k}$

{update next column of A}

end
end

When we study this algorithm, it may be clear that the choice $\lambda_{kk} = \nu_{kk} = \delta_k^{-1}$ gives optimal results, since it makes the calculations of $\bar{\ell}_{.k}$ and ν_{kj} trivial. The description of the algorithm then becomes much shorter, especially if the introduction of the names for L, D and U is eliminated:

for k = 1(1)n do
begin {maximum search and column interchange}
for j = k+1(1)n do
$a_{.j} \leftarrow a_{.j} - (\alpha_{kj}/\alpha_{kk})\bar{a}_{.k}$
end

On a 1-pipe CYBER 205, Hoffmann obtained the following MFLOPS-speeds for various values of the order n of the matrix A:

	n = 25	50	100	200	400
speed in MFLOPS	7.3	15.5	28.4	46.1	63.6

In Hoffmann (1985) many more experiments are reported and a comparison is made with standard routines for solving dense linear systems from the program libraries LINPACK, NAG and QQLIB. The above algorithm gives the smallest CPU-time.

Dongarra and Hewitt (1985) have implemented dense linear algebra algorithms on a CRAY X-MP-4 using multitasking and obtained a MFLOPS-speed of more than 700. They remark that a system of equations of order 1000 can now be factored and solved in less than one second!

3.2. Expressions: Evaluating a Polynomial

Given a real number x_0 and a polynomial

$$P(x) = a_0 + a_1x + a_2x^2 + \dots + a_nx^n,$$

the well-known rule of Horner for computing $P(x_0)$ reads as follows:

$$b_n := a_n ,$$

$$b_j := a_j + x_0 b_{j+1}, \quad j = n-1\ (-1)\ 0,$$

$P(x_0) := b_0$.

If we would have 2 processors, able to work in parallel, we could write P(x) as

$$P(x) = a_0 + a_2x^2 + a_4x^4 + \dots \qquad \text{(even powers)}$$
$$+ x(a_1 + a_3x^2 + \dots) \qquad \text{(odd powers)}$$

The first and the second processor could then compute the even and the odd powers sum, respectively, as follows:

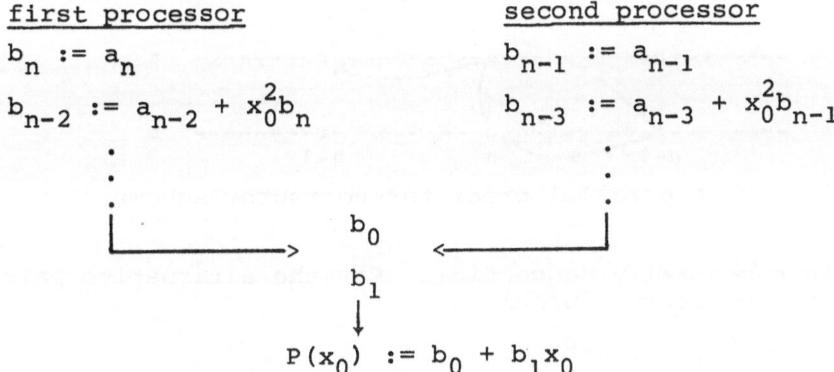

first processor
$b_n := a_n$
$b_{n-2} := a_{n-2} + x_0^2 b_n$
second processor
$b_{n-1} := a_{n-1}$
$b_{n-3} := a_{n-3} + x_0^2 b_{n-1}$
b_0
b_1
$P(x_0) := b_0 + b_1x_0$

This process can be generalized for many-processor systems.

3.3. ODEs

Let us consider the scalar ordinary differential equation

$$y' = f(x,y), \quad x > 0, \quad y(0) = y_0.$$

At first sight, it seems that there is little scope for parallelism in solving (scalar) ODEs, since the usual integration methods are essentially sequential. However, there exist parallel versions of serial predictor-corrector methods (cf. Miranker and Liniger, 1967). We will describe one of them in some detail. Fix a mesh size h and let $x_n := (n-1)h$, $n = 1,2,\dots,$ and let y_n be an approximation to the solution y at x_n. Then one serial predictor-corrector scheme is the following:

$$y_{n+1}^p := y_n^c + (h/2)[3f_n^c - f_{n-1}^c],$$

$$y_{n+1}^c := y_n^c + (h/2)[\,f_{n+1}^p + f_n^c],$$

where y_n^p and y_n^c are predicted and corrected values of y_n, respectively, and f_n^c and f_n^p represent $f(x_n,y_n^c)$ and $f(x_n,y_n^p)$, respectively. The sequel of computations is shown in Figure 2a where the upper line represents the process for y_n^p and f_n^p and the lower line for y_n^c and f_n^c. The sequence of computations here is: --> y_{n+1}^p --> f_{n+1}^p --> y_{n+1}^c --> f_{n+1}^c --> and the computational front is indicated by the dotted line. This

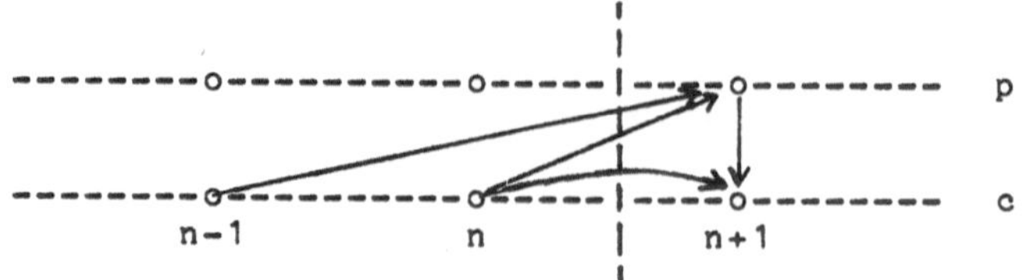

Fig. 2a. A serial predictor-corrector scheme.

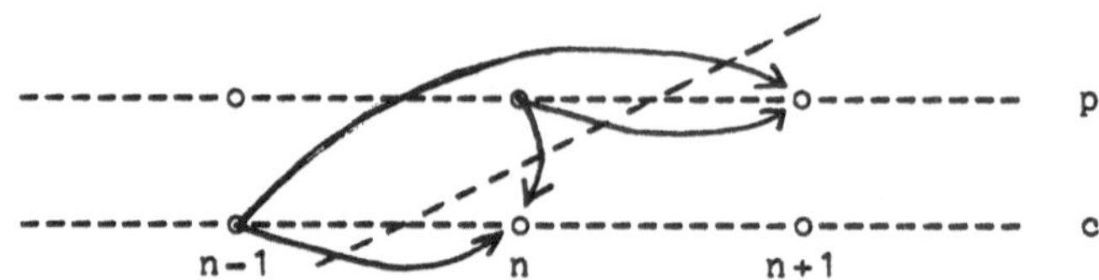

Fig. 2b. A parallel predictor-corrector scheme.

process is essentially sequential. For the alternative pair of predictor-corrector formulas

$$y^p_{n+1} := y^c_{n-1} + 2\, hf^p_n,$$

$$y^c_n := y^c_{n-1} + (h/2)[\, f^p_n + f^c_{n-1}],$$

the computational process may be divided into two concurrent parts:

$$\text{-->}\ y^p_{n+1}\ \text{-->}\ f^p_{n+1}\ \text{-->}$$

$$\text{-->}\ y^c_n\ \text{-->}\ f^c_n\ \text{-->}$$

which can be processed in parallel, since the computational front is now skewed (see Figure 2b). This kind of parallelization has been extended to many (≥ 2) processors and to other algorithms like the Runge-Kutta method.

3.4. PDEs

There is an extensive survey paper by Ortega and Voigt (1985) which surveys the present status of numerical methods for partial differential equations on vector and parallel processors, together with a discussion of applications in fluid dynamics (Navier-Stokes equations, potential equation, reservoir simulation, numerical weather prediction), structural analysis, acoustic wave propagation, plasma physics, design of VLSI devices, molecular dynamics, etc.

Generally spoken, discretization of place variables in PDEs may yield large systems of ODEs or of (non)linear equations. For vector and parallel machines, it seems that explicit methods for the former and iterative methods for the latter are more suitable than their respective counterparts: implicit and direct methods.

We mention here a few groups of people working on "vector and parallel software:

Schönauer: 3D problems;

Barkai, Brandt Hackbusch, Trottenberg, Stuben Hemker, Wesseling	multigrid methods;

Stelling, Wubs: shallow water equations;

etc. etc.

3.5. FFT

The discrete Fourier transform of a vector $a = (a_j)$, $j = 0,1,\ldots,N-1$, is given by

$$b_i = \sum_{j=0}^{N-1} \omega^{ij} a_j, \; i = 0(1)N-1, \; \omega = \exp(2\pi\, i/N).$$

This is a complex matrix-vector multiplication requiring $O(N^2)$ operations. However, in the Fast Fourier Transform the matrix $\Omega = (\omega^{ij})$ is factorized in (assuming, for simplicity, $N = 2^{N+1}$) $\log N = n + 1$ very simple matrices and the cumulative product is computed. The number of operations required now is $O(N \log N)$. The FFT algorithm can be described as follows:

for $r = 0(1)N-1$, $k = 0(1)n$, let

$$r := [r_0 r_1 \ldots r_n] = \sum_{j=0}^{n} r_{n-j}\, 2^j, \; r_i = 0 \text{ or } 1,$$

$$f(r,k) := [r_0 \ldots r_{k-1}\; 0\; r_{k+1} \ldots r_n],$$

$$h(r,k) := [r_0 \ldots r_{k-1}\; 1\; r_{k+1} \ldots r_n],$$

$$g(r,k) := [r_k r_{k-1} \ldots r_0\; 0 \ldots 0],$$

$$\text{rev}(r) := [r_n r_{n-1} \ldots r_0] \; (= g(r,n)).$$

FFT:

$z_i := \omega^i \; (i = 0(1)N-1)$,

$c_i := a_i \; (i = 0(1)N-1)$,

for $k := 0$ step 1 until n do

$c_i := c_{f(i,k)} + z_{g(i,k)} \times c_{h(i,k)} \; (i = 0(1)N-1)$,

od,

$b_i := c_{\text{rev}(i)} \quad (i = 0(1)N-1)$,

It should be observed that either $f(r,k) = r$ and $h(r,k) = r + 2^{n-k}$ or $f(r,k) = r - 2^{n-k}$ and $h(r,k) = r$, so the movements for c within the do-loop are well structured. One should be careful in order to avoid memory bank conflicts. The book by Hockney and Jesshope (1981) provides an excellent discussion of parallel aspects of the FFT and of other discrete transforms. Recent work on vectorizing the FFT (Fornberg, 1981; Korn and Lambiotte, 1979; Swartztrauber, 1984; Wang, 1980) indicates that the efficiency increases with the number of transforms.

3.6. Number Theory

In the last decade, methods for factorization of positive integers have attracted much attention, partly because of the discovery that the security of certain cryptographic systems depends on the difficulty of the decomposition of integers into prime factors (cf. Riesel, 1985). Factorization methods like the quadratic sieve method and Lenstra's recent elliptic curve factorization method have certain features by which these algorithms may be very suitable for implementation on vector and parallel computers. As an example, we mention one of the steps in the quadratic sieve methods, viz., to compute, modulo a given number N, the product of a large number M of integers with values between 0 and N-1. The scalar FORTRAN version of this step reads as follows: (here, it is assumed that the square of N can still be represented as an integer, and the integers to be multiplied are stored in the array A)

```
      INTEGER N, M, PROD, A(M)
      PROD = 1
      DO 1 I = 1, M
         PROD = MOD( PROD * A(I), N)
    1 CONTINUE
```

A vector version of this step has been implemented on a 1-pipe CYBER 205 and looks as follows:

```
      INTEGER N, M, PROD, A(M), B(M/2), C(M/2), K, K2
      REAL    REVN
      REVN = 1.0 / N
      K = M
    1 K2 = K / 2
      B(1; K2) = A(1; K2)  * A(K2+1; K2)
      C(1; K2) = B(1; K2)  * REVN
      A(1; K2) = B(1; K2)  - N * VINT(C(1;K2); K2)
      IF (2*K2 .EQ. K) THEN
         K = K2
      ELSE
         A(K2+1) = A(K)
         K = K2 + 1
      END IF
      IF (K .GT. 1) GOTO 1
      PROD = A(1)
```

The technique used here is a form of recursive doubling (cf. Section 4.1): two vectors which consist of the first and the second half part of A are multiplied (modulo N); the result is stored in the first half of A. This multiplication and storage step is repeatedly applied on A, where in each step the length of A is halved. The REAL REVN (= 1/N) is used because vector multiplication is much cheaper than vector division (on a 1-pipe CYBER 205, multiplication of two vectors of length N requires 52 + N clock cycles and division 80 +

25N/4). Vector syntax is used. E.g. A(I;J) is the vector consisting of A(I), A(I+1), ..., A(I+J-1). The vector function VINT computes the integral parts of all the elements of the vector argument. Some timings are given below.

N	scalar version	vector version
10,000	0.016 sec.	0.002 sec.
50,000		0.009 sec. (28 MFLOPS)

3.7. Numerical Verification of the Riemann Hypothesis

The Riemann hypothesis says that all the complex zeros of the Riemann zeta function $\zeta(s)$ have real part 1/2. This is a famous 125 year old statement of Riemann, which has resisted up to now the efforts of the best mathematicians to prove or disprove it. In order to verify the Riemann hypothesis numerically, it is necessary to know, for many (hundreds of millions) values of t, the sign of the following real-valued function:

$$Z(t) = 2 \sum_{k=1}^{m} k^{-\frac{1}{2}} \cos[t.\log(k) - \theta(t)] + R_m(t),$$

where $m = [(t/2\pi)^{\frac{1}{2}}]$. The time needed to compute $\theta(t)$ and $R_m(t)$ is negligible compared to the total time for Z(t).

Three versions to compute Z(t) on a 1-pipe CYBER 205 have been developed: a half, a normal and a double precision version. With the first very fast version about 99% of the values of Z(t) could be computed with certainty. With the second version, about 99% of the remaining values could be determined with certainty. The double precision version was accurate enough to cover all the remaining values. The half precision version gained a speed of about 134 MOPS (Million Operations per Second), the normal precision version about 57 MOPS, so that the CYBER 205 turned out to be extremely suitable to solve this problem (reasons: pipelining, different precisions possible, possibility to link operations, e.g. constructs like |A(I)|*B + |C(I)| require 1 clock cycle in a vector call on the CYBER 205; this corresponds to a speed of 400 million operations per second!). Details may be found in te Riele et al. (1985), Winter and te Riele (1985) and van de Lune et al. (1986).

4. VECTORIZATION AND PARALLELIZATION TECHNIQUES

Quite a number of techniques are known for generating parallel algorithms. One important distinction should be made in this respect: the number of available processors is limited or not. The latter case is interesting from a theoretical point of view, yielding results like: a non-singular n×n matrix can be inverted in $O(\log^2 n)$ time, using $O(n^4)$ processors (Csanky, 1976). The former case is more practical, since it is usually concerned with a particular processor with a given number of processing elements, or a pipelined processor with fixed characteristics like clock cycle time, start up time, memory bank cycle time.

In this Section examples of recursive doubling and cyclic reduction techniques will be treated. Moreover, techniques for matrix-vector and matrix-matrix multiplication will be discussed. Finally, some results will be given of implementation and optimization on a CYBER 205 of a set of standard matrix-vector subroutines (so-called Extended BLAS).

A number of examples given in this Section are derived from a lecture by H.A. van der Vorst presented at the Colloquium 'Numerical Aspects of Vector and Parallel Processing' at the meeting of September 27, 1985, which was held in Amsterdam.

4.1. Recursive Doubling

Recursive doubling is a powerful method of generating parallel algorithms. The basic idea is to repeatedly separate each computation into two independent parts of equal complexity which can then be computed in parallel. For example,

$$\sum_{i=1}^{N} a_i = (\sum_{i=1}^{n-1} a_i) + (\sum_{i=n}^{N} a_i), \quad n = \lceil N/2 \rceil,$$

and by further application of this splitting, the sum can be computed in $\lceil \log N \rceil$ steps using N/2 processors. This may be implemented on a pipelined processor as follows (here, like in Section 3.6, we use vector-syntax for the CYBER 205):

```
WHILE N > 1 DO
   M = ( N+1 ) / 2
   A( 1; M ) = A( 1; M ) + A ( M+1; N-M )
   N = M
OD
```

If an addition of two vectors of length N on a pipelined computer takes a+bN clock cycles (a is the start-up time), and if scalar addition takes c clock cycles, then the times needed for the sequential algorithm and for the parallel version are approximately cN and $a.\log_2 N + bN$ cycles. Comparing these two times, we can compute the approximate turning point for which the parallel version becomes faster than the sequential. Some examples are given in Table 5.

Table 5. Turning point from which parallel addition algorithm runs faster than sequential version.

a + bN: number of clock cycles needed for vector addition (length N)
c : number of clock cycles needed to add two scalars

Computer	a	b	c	value of N for which $a.\log_2 N + bN \approx cN$
CRAY-1	30	3	6	59
p-pipe CYBER 205	51	1/p	5	81 for p=1, 69 for p=2, 65 for p=4
CRAY X-MP	30	1		
FUJITSU VP-100	30	2		

Recursive doubling is applicable in a large number of instances. Table 6 is taken from Stone (1973). Theoretically, most of the recurrences mentioned there can be computed in O(log N) time if O(N) processors are available. However, actual implementation (like the one given above) is needed to show the real benefit obtainable with this technique.

Another example of recursive doubling occurs in the solution of bidiagonal linear systems Ax = b, where

$$A = \begin{bmatrix} 1 & & & & \\ a_2 & 1 & & & \\ & a_3 & 1 & & \\ & & \ddots & \ddots & \\ & & & a_N & 1 \end{bmatrix}, \quad x = \begin{bmatrix} x_1 \\ x_2 \\ x_3 \\ \vdots \\ x_N \end{bmatrix}, \quad b = \begin{bmatrix} b_1 \\ b_2 \\ b_3 \\ \vdots \\ b_N \end{bmatrix}$$

Table 6. Functions suitable for recursive doubling.

Function	Description
$X_i = X_{i-1} + a_i$	Sum the elements of a vector
$X_i = X_{i-1} * a_i$	Multiply the elements of a vector
$X_i = \min(X_{i-1}, a_i)$	Find the minimum
$X_i = \max(X_{i-1}, a_i)$	Find the maximum
$X_i = a_i X_{i-1} + b_i$	First order linear recurrence, inhomogeneous
$X_i = a_i X_{i-1} + b_i X_{i-2}$	Second order linear recurrence
$X_i = a_i X_{i-1} + b_i X_{i-2} + \ldots$	Any order linear recurrence, homogeneous or inhomogeneous
$X_i = (a_i X_{i-1} + b_i)/(c_i X_{i-1} + d_i)$	First order rational fraction recurrence
$X_i = a_i + b_i/X_{i-1}$	Special case of first order rational fraction
$X_i = \mathrm{sqrt}(X_{i-1}^2 + a_i^2)$	Vector norm

The standard solution method is:

$$x_1 := b_1$$

$$x_i := b_i - a_i * x_{i-1},\ i = 2,3,\ldots,N.$$

Some (scalar) improvement can be obtained by loop-unrolling (cf. van der Vorst and van Kats, 1984, section 3.8). A recursive doubling technique for solving the bidiagonal linear system can be described as follows. Left-multiplication by the matrix $-A + 2I$ yields the equation $A'x = b'$ where

$$A' = \begin{bmatrix} 1 & & & & & \\ 0 & 1 & & & & \\ a_3' & 0 & 1 & & & \\ & a_4' & 0 & 1 & & \\ & & \cdot & \cdot & \cdot & \\ & & & \cdot & \cdot & \cdot \\ & & & a_N' & 0 & 1 \end{bmatrix},$$

with obvious values of a_i' $(i = 3,4,\ldots,N)$ and b_i' $(i = 1,2,\ldots,N)$. Next, we left-multiply with the matrix $-A' + 2I$ to obtain the equation $A''x = b''$, where

$$A'' = \begin{bmatrix} 1 & & & & & & & \\ 0 & 1 & & & & & & \\ 0 & 0 & 1 & & & & & \\ 0 & 0 & 0 & 1 & & & & \\ a_5'' & 0 & 0 & 0 & 1 & & & \\ & a_6'' & 0 & 0 & 0 & 1 & & \\ & & a_7'' & 0 & 0 & 0 & 1 & \\ & & & \cdot & \cdot & \cdot & \cdot & \cdot \\ & & & & \cdot & \cdot & \cdot & \cdot \end{bmatrix}$$

Repeating this process at most $\lceil \log_2 N \rceil$ steps eliminates all the unknowns and yields the solution in the vector b. On a parallel processor with N processing elements, this would yield the solution in about $\log_2 N$ time-steps. However, the total number of operations in this parallel algorithm is much larger than that in the serial version and, even though the operations now are all vector operations, the actual performance on a vector computer (like the CRAY-1) is worse than the loop-unrolled version. Speeds of 2-5 MFLOPS are reported for the parallel version implemented on pipelined computers. For the loop-unrolled scalar version, van der Vorst and van Kats (1984) report speeds of 8.0 MFLOPS on a CRAY-1 and 7.8 on a CYBER 205 (1- and 2-pipe) with N = 5000. Very recently, J. Schlichting of CDC and the CWI managed to reach a speed of 12 MFLOPS with N = 3000.

4.2. Cyclic Reduction

It seems that the technique of cyclic reduction was first used to solve tridiagonal equations by Hockney in 1965 (in collaboration with Golub; cf Hockney and Jesshope, 1981). We illustrate this technique with the bidiagonal equation of the previous section. Eliminating x_1 from the second equation, x_3 from the fourth, and so on, we obtain the system

$$\begin{bmatrix} 1 & & & \\ a'_4 & 1 & & \\ & a'_6 & 1 & \\ & & \ddots & \ddots \end{bmatrix} \begin{bmatrix} x_2 \\ x_4 \\ x_6 \\ \vdots \end{bmatrix} = \begin{bmatrix} b'_2 \\ b'_4 \\ b'_6 \\ \vdots \end{bmatrix},$$

where $a'_{2i} = -a_{2i} * a_{2i-1}$, $b'_{2i} = b_{2i} - a_{2i} * x_{2i-1}$. This process can be repeated, if suitable, and on a parallel processor with N processors, this algorithm needs about $\log_2 N$ steps. In van der Vorst and van Kats (1984) experiments with this algorithm on pipelined computers like the CRAY-1 and the CYBER 205 are reported. On the CRAY-1 speeds close to 12 MFLOPS (for N = 5000) were obtained, and 9 MFLOPS on a 2-pipe CYBER 205.

4.3. Matrix-Vector and Matrix-Matrix Multiplication

The usual method of computing the matrix-vector product y := Ax is by taking inner products of rows of A with x:

$$y_i := \sum_{j=1}^{N} a_{ij} x_j, \quad i = 1,2,\ldots,N.$$

Implementing this method on a CRAY-1, a speed of 53 MFLOPS is obtained (N = 300). For N = 200 and 201, the speeds are 37 and 49 MFLOPS, respectively (this is due to memory bank conflicts). On the CYBER 205, however, there is a so-called stride problem which means that elements have to be loaded from memory which are not stored in continuous locations (arrays in Fortran are stored column-wise). The speed obtained on a 1-pipe CYBER 205 is 5.7 MFLOPS. However, by reordering the computations column-wise:

$$y := x_1 a_{.1} + x_2 a_{.2} + \ldots + x_N a_{.N},$$

where $a_{.i}$ is the i-th column of the matrix A, the speed obtained on a 1-pipe CYBER 205 is 66 MFLOPS and on a 2-pipe CYBER 205 106 MFLOPS! If, however, we would have to compute y := A'x, the inner product version should be used on the CYBER 205. If the matrix is symmetric and if only the upper (or lower) part is available in storage, a combination of the two methods mentioned above should be used on the CYBER 205.

For band matrices, the picture is quite different (cf. Madsen et al., 1976). Consider, for example, the tridiagonal band-matrix product

$$\begin{bmatrix} y_1 \\ y_2 \\ y_3 \\ \cdot \\ \cdot \\ \cdot \\ y_N \end{bmatrix} := \begin{bmatrix} a_{11} & a_{12} & & & & \\ a_{21} & a_{22} & a_{23} & & & \\ & a_{32} & a_{33} & a_{34} & & \\ & & \cdot & \cdot & \cdot & \\ & & & \cdot & \cdot & \cdot \\ & & & & \cdot & \cdot \quad \cdot \\ & & & & a_{N,N-1} & a_{NN} \end{bmatrix} \begin{bmatrix} x_1 \\ x_2 \\ x_3 \\ \cdot \\ \cdot \\ \cdot \\ x_N \end{bmatrix} .$$

In order to save space, band matrices are usually stored in rectangular arrays such that the non-zero diagonals are stored in rows or columns of the array. The above multiplication can be executed very efficiently on a vector computer by expressing the product as a sum of three vector-vector multiplications:

$$\begin{bmatrix} y_1 \\ y_2 \\ y_3 \\ \cdot \\ \cdot \\ \cdot \\ y_{N-1} \\ y_N \end{bmatrix} := \begin{bmatrix} \\ a_{21} \\ a_{32} \\ \cdot \\ \cdot \\ \cdot \\ a_{N-1,N-2} \\ a_{N,N-1} \end{bmatrix} * \begin{bmatrix} \\ x_1 \\ x_2 \\ \cdot \\ \cdot \\ \cdot \\ x_{N-2} \\ x_{N-1} \end{bmatrix} + \begin{bmatrix} a_{11} \\ a_{22} \\ a_{33} \\ \cdot \\ \cdot \\ \cdot \\ a_{N-1,N-1} \\ a_{NN} \end{bmatrix} * \begin{bmatrix} x_1 \\ x_2 \\ x_3 \\ \cdot \\ \cdot \\ \cdot \\ x_{N-1} \\ x_N \end{bmatrix} + \begin{bmatrix} a_{12} \\ a_{23} \\ a_{34} \\ \cdot \\ \cdot \\ \cdot \\ a_{N-1,N} \end{bmatrix} * \begin{bmatrix} x_2 \\ x_3 \\ x_4 \\ \cdot \\ \cdot \\ \cdot \\ x_N \end{bmatrix}$$

Hence, the time for this expression is essentially 5N clock cycles (3N for the multiplication and 2N for the addition of the vectors), if we neglect start-up times and if the diagonals of the matrix are stored column-wise. If there are diagonals with constant value, the number of clock cycles can be decreased by N for each constant diagonal (except the last) by using chaining or linked triads.

The ideas described here can be extended to matrix-matrix multiplication. For example, in Dongarra et al. (1984) six possible permutations of the three loop indices in matrix-matrix multiplication programming are described. This gives rise to six possible implementations of matrix multiplication. Each implementation has quite different memory access patterns, which will have an important impact on the performance of the algorithms on a given vector or parallel processor (also cf. Madsen et al., 1976).

4.4. Extended BLAS

Recently, Dongarra et al. (1984) have proposed a standard set of routines for matrix-vector multiplication, rank-1 and rank-2 updates, and inversion of triangular systems of equations called the set of extended BLAS (Basic Linear Algebra Subroutines). This extends the existing set of BLAS, which are standard routines for operations on vectors (Lawson et al., 1979). The extended BLAS routines will become available in

Fortran 77. Besides that the proposers hope that efficient implementations will become available on a wide range of computer architectures. At the CWI an implementation of the extended BLAS on the CYBER 205 is being developed. Here we will give a short description of the contents of the set of extended BLAS, and present some timings obtained on the CYBER 205.

Three types of basic operations (MV, R1/R2 and TR) are proposed:

a) Matrix vector (MV) products of the form

$$y := \alpha Ax + y, \text{ and } y := \alpha A'x + y$$

where α is a scalar, x and y are vectors and A is a matrix. and

$$x := Tx \text{ and } x := T'x$$

where T is an upper or lower triangular matrix.

b) Rank-one (R1) and rank-two (R2) updates of the form

$$A := \alpha xy' + A \quad \text{and} \quad A := \alpha xy' + \alpha yx' + A.$$

c) Solution of triangular equations (IV) of the form

$$x := T^{-1} x,$$

where T is an upper or lower non-singular triangular matrix. The subroutines have a name which consists of five characters. The first character is an S (indicating real versions; other possibilities are C for complex, D for double precision). Characters two and three denote the kind of matrix involved and the final two characters denote the type of operations. There are sixteen subroutines, marked by an * below.

type of matrix	operation MV	R1	R2	IV
GE general matrix	*	*		
GB general band	*			
SY symmetric	*	*	*	
SP symmetric packed	*	*	*	
SB symmetric band	*			
TR triangular	*			*
TP triangular packed	*			*
TB triangular band	*			*

Table 7 gives timings in milliseconds of the 16 subroutines. The timings of the packed matrix versions are the same as those of the unpacked versions, and are omitted. Two timings per routine are given: that of the Fortran 77 version and that of the 1-pipe CYBER 205 optimized version. The final column gives MFLOPS-speeds of the optimized versions. The order of the matrices used was 500 for full matrices and 30000 for band matrices. In the case of band matrices, the number of non-zero diagonals was: 2 upper and 3 lower in the general case, 2 in the symmetric case and 5 in the triangular case.

Table 7. Timings (in msec.) of Extended BLAS subroutines.

Subroutine	Fortran 77	CYBER 205 Optimized	MFLOPS-speed of opt.
SGEMV	7	7	75
SGBMV	107	15	26
SSYMV	58	8	63
SSBMV	184	12	27
STRMV	4	4	63
STBMV	98	14	24
SGER1	7	7	75
SSYR1	4	4	60
SSYR2	8.5	7.8	64
STRIV		6	62
STBIV		230	2

The general matrix routines all run with a speed which comes quite close to the optimal speed of 100 MFLOPS, obtainable for general matrix multiplication. In the band matrix case, the speeds are negatively influenced by the row-wise storage convention for the diagonals: this requires gathering of the diagonal elements. If column-wise storage would be allowed, then the MFLOPS-speeds could be multiplied by a factor of at least 1.6, which would bring these speeds reasonably close to the optimum of 50 MFLOPS, obtainable for band matrix operations.

5. REFERENCES

Bernutat-Buchmann, U., Rudolph, D., and Schlosser, K.-H., 1983, "Parallel Computing I, Eine Bibliographie", Bochumer Schriften zur Parallelen Datenverarbeitung 1, Computer Centre, Bochum, second ed., September 1983.

Böhm, A.P.W., 1983, "Dataflow Computation", CWI Tract 6, Centre for Mathematics and Computer Science, Amsterdam.

Csanky, L., 1976, Fast parallel matrix inversion algorithms, SIAM J. Comput. 5:618-623.

Dongarra, J.J., Du Croz, J., Hammarling, S., and Hanson, R.J., 1984, A proposal for an extended set of Fortran Basic Linear Algebra Subprograms, Techn. Memo. No. 41, Argonne National Lab., Argonne, Ill. 60439, December 1984.

Dongarra, J.J., Gustavson, F.G., and Karp, A., 1984, Implementing linear algebra algorithms for dense matrices on a vector pipeline machine, SIAM Review 26:91-112.

Dongarra, J.J., and Hewitt, T., 1985, Implementing dense linear algebra algorithms using multitasking on the CRAY X-MP-4 (or approaching the gigaflop), Techn. Memo. No. 55, Argonne National Lab., Argonne, August 1985.

Duff, I.S., and Reid, J.K., eds., 1985, "Vector and Parallel Processors in Computational Science", proceedings of the Second International Conference in VPPCS, Oxford, August 1984, North-Holland, Amsterdam.

Emmen, A.H.L., ed., 1985, "Supercomputer Applications", Proceedings of the International Supercomputer Applications Symposium, Amsterdam, November 1984, North-Holland, Amsterdam.

Flynn, M.J., 1966, Very high speed computing systems, Proc. IEEE 14:1901-1909.

Fornberg, B., 1981, A vector implementation of the Fast Fourier Transform Algorithm, Math. Comp. 36:189-191.

Heller, D., 1978, A survey of parallel algorithms in numerical linear algebra, SIAM Review 20:740-777.

Hockney, R.W., and Jesshope, C.R., 1981, "Parallel Computers", Adam Hilger Ltd., Bristol.

Hoffmann, W., 1985, Gaussian elimination algorithms on a vector computer, Rept. 85-10, Dept. of Math., Univ. of Amsterdam, June 1985.

Kindervater, G.A.P., and Lenstra, J.K., 1983, "Parallel algorithms in Combinatorial Optimization: an Annotated Bibliography", Report BW189/83, Centre for Mathematics and Computer Science, August 1983.

Korn, D.G., and Lambiotte, J.L., Jr., 1979, Computing the Fast Fourier Transform on a vector computer, Math. Comp. 33: 977-992.

Lawson, C., Hanson, R., Kincaid, D., and Krogh, F., 1979, Basic linear algebra subprograms for Fortran usage, ACM Trans. on Math. Software 5:308-323.

Madsen, N.K., Rodrigue, G.H., and Karush, J.I., 1976, Matrix multiplication by diagonals on a vector/parallel processor, Information Process. Letters 5:41-45.

Miranker, W.L., 1971, A survey of parallelism in numerical analysis, SIAM Review 13:524-547.

Miranker, W.L., and Liniger, W.M., 1967, Parallel methods for the numerical integration of ODEs, Math. Comp. 21:303-320.

Ortega, J.M., and Voigt, R.G., 1985,"Solution of Partial Differential Equations on Vector and Parallel Computers", ICASE Rept. No. 85-1, NASA Langley Research Center, Hamption, Virginia 23665, January 1985.

Riesel, H., 1985, "Prime Numbers and Computer Factorization", Birkhauser.

Rodrigue, G.H., Madsen, N.K., and Karush, J.I., 1979, Odd-even reduction for banded linear equations, JACM 26:72-81.

Sameh, A.H., 1977, Numerical parallel algorithms - A survey, in: "High Speed Computer and Algorithm Organization", D.J. Kuck et al., eds., Academic Press, p. 207-228.

Schendel, U., 1984, "Introduction to Numerical Methods for Parallel Computers" (translated from German), Ellis Horwood Ltd., Chichester.

Schwartz, J., 1983, "A Taxonomic Table of Parallel Computers, based on 55 Designs", Ultra Computer Note No. 69, Courant Institute, New York University.

Shore, J.E., 1973, Second thoughts on parallel processing, Comp. Elect. Eng. 1:95-109.

Stone, H.S., 1973, An efficient parallel algorithm for the solution of a tridiagonal linear system of equations, JACM 20:27-38.

Stone, H.S., 1973, Problems of parallel computation, in: "Complexity of Sequential and Parallel Numerical Algorithms", J.F. Traub, ed., Academic Press, p. 1-16.

Swartztrauber, 1984, FFT algorithms for vector computers, Parallel Computing 1:45-63.

Sweet, R.A., 1977, A cyclic reduction algorithm for solving block-tridiagonal systems of arbitrary dimension, SIAM J. Numer. Anal. 14:706-719.

te Riele, H.J.J., Winter, D.T., and van de Lune, J., 1985, Numerical verification of the Riemann hypothesis on the CYBER 205, in: "Supercomputer Applications", A.H.L. Emmen, ed., North-Holland, Amsterdam, p. 33-38.

Traub, J.F., ed., 1973, "Complexity of Sequential and Parallel Numerical Algorithms", Academic Press.

van de Lune, J., te Riele, H.J.J., and Winter, D.T.,1986, On the zeros of the Riemann zeta function in the critical strip. IV., Math. Comp. 46:667-681.

van der Vorst, H.A., and van Kats, J.M., 1984, "The Performance of Some Linear Algebra Algorithms in Fortran on CRAY-1 and CYBER 205 Supercomputers", Techn. Rep. TR-17, ACCU, Utrecht.

van Leeuwen, J., 1983, "Parallel Computers and Algorithms",Rept. RUU-CS-83-13, RU Utrecht, September 1983.

Wang, H.H., 1980, On vectorizing the Fast Fourier Transform, BIT 20:233-243.

Winter, D.T., and te Riele, H.J.J., 1985, Optimization of a program for the verification of the Riemann hypothesis, Supercomputer 5:29-32.

Zakharov, V., 1984, Parallelism and array processing, IEEE Trans. Comp. C-33:45-78.

VECTORIZATION OF LARGE QUANTUM CHEMICAL PROGRAMS: METHODS AND ILLUSTRATIONS

Walter Ravenek

Scheikundig Laboratorium der Vrije Universiteit
De Boelelaan 1083, NL-1081 HV Amsterdam, The Netherlands

INTRODUCTION

One of the major tasks of quantum chemistry is the computation of the electronic wave function of molecular systems. The approximation that plays a central role in the theory is the one-electron model in which each electron is assumed to move independently in an effective potential consisting of the potential of the nuclei in the system and an averaged potential due to the other electrons. Much of our understanding of the chemical bond can be derived from this model. Apart from being a conceptual tool, the results from calculations using the one-electron model serve as a starting point for more refined calculations of the electronic wave function, in which the effect of electron correlation is taken into account. In addition to this, the results of calculations with the one-electron model can also be used to study other problems, e.g. the molecular dynamics problem.

By far the most frequently used one-electron model is the Hartree-Fock model. It has been implemented in various standard packages, which are routinely used by large numbers of quantum chemists and also be more experimentally oriented scientists. In spite of the conceptual simplicity of the underlying model, the programs themselves are not at all simple: they are technically quite complicated, and they make large demands on the resources of the computer system used. Large numbers of integrals need to be handled on each cycle of an iterative procedure. The computation time required rises as the fourth power of the size of the molecular system.

The computational complexity of methods that go beyond the one-electron model is even larger. This limits the size of the systems that can be dealt with to such an extent that many chemically interesting systems can only be treated at a relatively low level of sophistication. For example, the German quantum chemist Ahlrichs recent performed calculations on the molecules P_2, P_4 and P_8 on the 2 pipe Cyber 205 at Karlsruhe [1]. P_2 could be dealt with very well: a Hartree-Fock calculation and extensive configuration interaction (CI) to introduce electron correlation. For P_4 a good Hartree-Fock calculation was performed, some correlation was introduced by CI. For P_8 only a moderately accurate Hartree-Fock could be performed. These calculations were not just done with a standard program but with a package that was partially vectorized. Evidently the use of supercomputers alone is not sufficient to solve the computational problems of quantum chemistry.

In this paper we restrict ourselves to one-electron methods. In Section 2 we discuss two of these methods, viz. the Hartree-Fock and the Hartree-Fock-Slater-LCAO methods, concentrating on their computational aspects.

In Section 3 we deal with the strategy for the vectorization of large program packages. At our institute we are presently working on the vectorization of the HFS-LCAO method, a package of over 80,000 lines of Fortran. The supercomputer available to us is the Cyber 205 at SARA in Amsterdam. For these reasons our examples will be from this context. However, the principles outlined have wider applicability.

In Section 4 we discuss two examples in some detail, viz. the construction of the Fock matrix by numerical integration, as used in the HFS-LCAO method, and the matrix multiplication problem. These examples are meant to illustrate two aspects of vectorization: the analysis and the implementation of algorithms.

In Section 5, finally, we summarize our findings.

2. COMPUTATIONAL ASPECTS OF ONE-ELECTRON METHODS

In order to indicate which computational problems are involved in quantum chemistry, we will discuss two one-electron methods in some detail, viz. the Hartree-Fock (HF) and the Hartree-Fock-Slater (HFS)-LCAO methods. As mentioned in the introduction the purpose of these methods is the calculation of the electronic wave function Ψ of molecular systems. Quantum mechanics learns us that Ψ satisfies the Schrödinger equation

$$H\Psi = E\Psi \tag{2.1}$$

We limit ourselves to the nonrelativistic Schrödinger equation in the clamped nuclei (Born-Oppenheimer) approximation. In that case the Hamiltonian H is given by

$$\begin{aligned} H &= \sum_i \left(-1/2\ \Delta(i) - \sum_A \frac{Z_A}{R_{iA}}\right) + \sum_{i<j} \frac{1}{r_{ij}} \\ &= \sum_i h(i) + \sum_{i<j} g(i,j). \end{aligned} \tag{2.2}$$

Here h(i) is the one-electron operator consisting of the kinetic energy and the nucleus-electron interaction terms, and g(i,j) is the electron-electron interaction term. The lables i and j in Eq. (2.2) refer to electrons, the lable A refers to a nucleus.

2.1. The Hartree-Fock Method

In the derivation of closed-shell Hartree-Fock equations one determines the optimal wave function of the form

$$\Psi = (N!)^{-1/2} \det|\psi_1(1)\psi_1(2)\psi_2(3) \ldots \psi_{N/2}(N)|, \qquad (2.3)$$

where by convention only the diagonal of the matrix is given. One uses an expansion of the one-electron wave functions Ψ_a in a set of basis functions $\{\phi_\mu\}$,

$$\Psi_a = \Sigma_\mu \phi_\mu C_{\mu a} . \qquad (2.4)$$

The basis set expansion technique is used throughout quantum chemistry. It leads to a matrix formulation of the computational problem.

We need not go through the derivation here, it can be found in any text book on quantum chemistry [e.g. Ref. 2]. The final result is that the optimal matrix C with expansion coefficients satisfies the generalized eigenvalue problem

$$F\,C = S\,C\,\varepsilon, \qquad (2.5)$$

where the Fock matrix F is the matrix representation of the Fock operator

$$f(r) = h(r) + V_c(r) + V_x(r), \qquad (2.6)$$

afforded by the basis $\{\phi_\mu\}$:

$$F_{\mu\nu} = \int dr\, \phi_\mu(r)\, f(r)\, \phi_\nu(r). \qquad (2.7)$$

The one-electron operator h is defined in Eq. (2.2). The Coulomb operator V_c is given by

$$V_c(r_1) = \Sigma_{\sigma\lambda} P_{\sigma\lambda} \int dr_2\, \phi_\lambda(r_2) r_{12}^{-1}\, \phi_\sigma(r_2), \qquad (2.8)$$

and the exchange operator V_x by

$$V_x(r_1) = \Sigma_{\sigma\lambda} P_{\sigma\lambda} \int dr_2\, \phi_\lambda(r_2) r_{12}^{-1}\, \mathcal{P}_{12}\, \phi_\sigma(r_2), \qquad (2.9)$$

with $\mathcal{P}_{12}$ the operator that interchanges electrons 1 and 2. The matrix P occurring in Eqs. (2.8) and (2.9) is the so-called density matrix, whose elements are given by

$$P_{\mu\nu} = 2\, \Sigma_a C_{\mu a} C_{\nu a}, \qquad (2.10)$$

with the sum over the occupied orbitals in the determinant wave function (2.3). Furthermore, S is the overlap matrix of the basis $\{\phi_\mu\}$ and ε the diagonal matrix with eigenvalues.

We note the matrix elements of the Coulomb and the exchange operators in the basis $\{\phi_\mu\}$ can be written as sums over the two-electron integrals

$$\langle\mu\sigma|\nu\lambda\rangle = \int dr_1\, dr_2\, \phi_\mu(r_1)\, \phi_\sigma(r_2) r_{12}^{-1}\, \phi_\nu(r_1)\, \phi_\lambda(r_2) \quad (2.11)$$

weighted by elements of the density matrix P.

It can be seen that Eq. (2.5) is not really an eigenvalue problem, since the required coefficients C are implicitly contained in the Fock matrix F, through the density matrix P. The equation can be solved by use of iteration: the Fock matrix is calculated from a start density; solving the eigenvalue problem we calculate a new density, which can be used to calculate the Fock matrix, and so on. If the process converges (and it usually does) we finally obtain an electron distribution that is consistent with the potential field which it generates itself. For this reason the procedure for solving the Hartree-Fock equation is referred to as the Self-Consistent-Field method.

In Figure 1 we have summarized the computational procedure of the Hartree-Fock method. Steps 4, 5 and 6 form the iterative procedure.

We now turn to the question of vectorizing the Hartree-Fock method. It is becoming clear from attempts of scientists from various disciplines that the vectorization of computational methods is usually very easy or very hard, there being only a small middle ground. The vectorization of the Hartree-Fock method is an example of a hard problem. All problems essentially derive from the occurrence of the two-electron integrals $\langle\mu\sigma|\nu\lambda\rangle$. Taking into account the symmetry properties of the general term it can be easily shown that there are roughly 1/8 N^4 unique two-electron integrals, when the size of the basis $\{\phi_\mu\}$ is N. A basis set of 100 functions is not exceptionally large, in which case one already has 12.5 10^6 integrals to deal with. The calculation of these integrals is hard to vectorize. Furthermore, the integrals have to be manipulated in each cycle of the iterative procedure. Fortunately, many of the two-electron integrals are almost or exactly zero (typically 80%), so that they may be ignored. At this point we enter the field of sparse matrices, however, and operations on sparse matrices are much harder to vectorize than operations on dense matrices.

1) select a basis $\{\phi_\mu\}$
2) calculate integrals
   - one-electron
   - two-electron
3) guess start density matrix P
4) calculate F(P)
5) solve eigenvalue problem
6) calculate P
7) calculate required properties

Fig. 1. Schematic representation of the computational procedure for the Hartree-Fock method.

In the Hartree-Fock method large numbers of integrals have to be calculated. On each cycle of the iterative procedure these integrals have to be read in from secondary storage and have to be processed using sparse matrix techniques. All of these are major problems. We may conclude that the central method in quantum chemistry is hard to vectorize.

2.2. The Hartree-Fock-Slater-LCAO Method

In this section we take a look at the Hartree-Fock-Slater (HFS) method. This method is a one-electron method, just as the Hartree-Fock method. We will deal with a particular implementation of the HFS method, called the HFS-LCAO method, developed at the Vrije Universiteit in Amsterdam by Baerends and co-workers [3].

There are differences between the HFS-LCAO method and the HF method both in physical model and in computational scheme. Physically the HFS method differs from the HF method in the form of the exchange potential. Instead of the exchange potential (2.9) the HFS method employs

$$V_x(r) = -c_x \, [\Sigma_{\sigma\lambda} \, P_{\sigma\lambda} \, \phi_\lambda(r) \, \phi_\sigma(r)]^{1/3}, \tag{2.12}$$

which is a local potential, i.e. as an operator V_x is a multiplicative term, whereas the exchange potential in the HF method is nonlocal. For this reason the HFS method falls into the class of the so-called local (spin) density functional methods. In fact, the HFS-LCAO program has the option of using various other exchange potentials, but the one in Eq. (2.12) is the one commonly used.

The fact that the exchange potential is local allows a number of computational techniques to be used which greatly reduce the amount of work to be performed in a calculation. There are two major points: the use of a numerical integration scheme to perform the three dimensional integration in the calculation of the elements of the Fock matrix,

$$\begin{aligned} F_{\mu\nu} &= \int dr \, \phi_\mu(r) \, f(r) \, \phi_\nu(r) \\ &= \Sigma_k \, w(r_k) \, \phi_\mu(r_k) \, f(r_k) \, \phi_\nu(r_k), \end{aligned} \tag{2.13}$$

and the expansion of the one-electron density ρ in a set of fit functions $\{f_i\}$,

$$\rho(r) = \Sigma_i \, a_i \, f_i(r), \tag{2.14}$$

which is used to simplify the calculation of the Coulomb potential:

$$\begin{aligned} V_c(r_1) &= \int dr_2 \, (r_{12})^{-1} \, \rho(r_2) \\ &= \Sigma_i \, a_i \int dr_2 \, (r_{12})^{-1} \, f_i(r_2). \end{aligned} \tag{2.15}$$

These features make the computational complexity of order N^3 instead of N^4 as for the Hartree-Fock method. It has been shown that the results obtained with the HFS-LCAO method agree at least as well with experiment as the corresponding results obtained with the Hartree-Fock method [4], but with less computational effort. Contrary to the Hartree-Fock method, however, it is hard to improve upon the results obtained in a systematic way.

As a final technical point we mention that the matrices that occur in the computational scheme of the HFS-LCAO method are dense, i.e. sparse matrix techniques are not needed. This greatly simplifies the job of vectorization.

In Figure 2 we give the computational scheme for the HFS-LCAO method. It can immediately be seen that the scheme is very similar to the one presented for the HF method. The main difference lies in the fact that the introduction of the fit basis causes a greater diversity in tasks to be performed.

As will be shown in Section 4 the numerical integration is particularly suited for vectorization. The construction of the Fock matrix normally takes 50-75% of the computation time. However, not only the construction of the Fock matrix is vectorizable, also the calculation of the integrals that are needed in the construction step can be easily vectorized.

Thus, the reasons for the suitability of the HFS-LCAO method for vectorization are a combination of all the essential features: the use of a local potential, the use of a numerical integration scheme for the construction of the Fock matrix, and the avoidance of two-electron integrals, keeping **all** matrices in the problem dense.

3. VECTORIZATION STRATEGY

In this section we discuss some aspects of the overall strategy in the vectorization of a program. We distinguish two aspects in the following, the vectorization of the algorithm on

1) select a basis $\{\phi_\mu\}$
   select a fit basis $\{f_i\}$
2) calculate integrals
   - kinetic energy
   - values basis functions
   - values fit functions
   - Coulomb integrals fit functions
   - other fit integrals
3) guess start density matrix P
4) calculate $f(r_k)$
   calculate F
5) solve eigenvalue problem
6) calculate P and density fit
7) calculate required properties

Fig. 2. Schematic representation of the computational procedure for the Hartree-Fock-Slater-LCAO method.

the one hand and the implementation of the vectorized algorithm on the other hand. Most of the things that will be said are well-known and completely general. At a number of points we mention some aspects of the vectorization of the HFS-LCAO package. In Section 4 we give some more detailed examples.

3.1. Vectorization of the Algorithm

The most important point seems so obvious that it hardly needs discussion: identify long vectors in the problem at hand. For the Cyber 205 they must be stored in contiguous memory locations. For computers that can handle strides greater than one efficiently, such as the Cray, it is not strictly necessary to deal with vectors stored this way, but it may be useful to do so as much as possible in order to avoid memory bank conflicts.

In some problems the identification of long vectors may be easy, in other problems a new point of view on the algorithm and the data structure may be required. Often the data structure has to be modified, which implies the rewriting of a part, possibly a large part, of the program. The latter is the case for the HFS-LCAO package.

It is possible to analyse the behaviour of a simple algorithm accurately on pipelined computers such as the Cyber 205. In order to predict the CPU time needed for an algorithm one can use the well-known relation for the time t(N) needed for a vector operation on vectors of length N [5]

$$t(N) = c(N_{1/2}+N), \tag{3.1}$$

with c and $N_{1/2}$ constants that depend on the vector operation. Relation (3.1) also holds in an approximate sense for data movement with gathers and scatters (deviations occurring because of memory bank conflicts). It may be noted that the relation can be used not only to predict but also to check afterwards if the code performs as expected.

Our next point is the I/O problem. Often it happens that a vectorized algorithm is completely I/O bound. In such a case it may make sense to recalculate certain quantities. In quantum chemistry there exist Hartree-Fock programs that recalculate large numbers of integrals on each cycle of the iterative procedure, even on scalar computers. If such a drastic measure is not possible or not effective one has to do the I/O. Obviously the efficient ways of performing I/O may be different on different computers. For a discussion pertaining to the Cyber 205 we refer to [6].

Also memory usage may become a bottleneck on supercomputers, even with virtual memory. There is a tendency for memory usage to become very large in vectorized programs. One might try to solve the I/O problem by declaring the files as very large matrices. Furthermore, there is the creeping danger that in the vectorization process scalars tend to become vectors, and vectors tend to become matrices, thereby increasing memory usage considerably. One may end with a program that spends most of its execution time doing page faults. Therefore it is worthwhile to reduce memory usage within reasonable limits.

Again, well-known techniques exist, e.g. storing only a triangle of a symmetric matrix, and overlaying results and data from which they were calculated.

3.2. Implementation of the Algorithm

Let us suppose that we are in the possession of a well-vectorized algorithm, that makes an intelligent use of I/O and memory. How do we implement it in a program? Also here a strategy is required. In my opinion this aspect of vectorization generally received too little attention. The race for results on a short notice seems to encourage us to ignore such matters.

This issue becomes all the more urgent if the program to be vectorized becomes larger. The HFS-LCAO package is large, it consists of over 80,000 lines of Fortran. In the vectorization of the package we have made transportability a central issue. In this section we discuss the reasons, the means and possible disadvantages of this strategy. Once again, much has been said and written on this issue [7]. We summarize the points that seem most relevant for our purpose.

The reasons are as follows:
- We want the package to have a long life. That is, we want it to survive new releases of the operating system, but also the computer on which we run the package.
- A large program forms a management problem in itself [8]. At the same time various people are working on the program, adding new possibilities, correcting mistakes. At our institute the HFS-LCAO package runs on three computers. It is highly undesirable to have more than one version of the package. Everyone who has worked in such an environment knows how strong the tendency is for different versions to grow further apart. Therefore we want to minimize the changes needed for going from one computer to the other.
- A large program implies a large investment of human labour. Other scientists should be able to use it. This will only happen if the program can be transported with a reasonable amount of work.

Our strategy to attain portability is the following:
- We use standard Fortran 77, that is, we do not use the vector extensions of "Fortran 200", available on the Cyber 205. The actual generation of vectorized code is left to the compiler, possibly in collaboration with a preprocessor (VAST [9]). This method is called automatic vectorization. We use it for the bulk of the code.
- In conjunction with the above strategy it is necessary to localize and modularize the critical parts of the program. In this respect we think of the interaction with the operating system in general, and the I/O in particular, and of the parts of the program that are computationally most expensive. For some of the critical parts we accept a nonstandard version, but we try to minimize their number.
- As a third point we mention the fact that we do not use routines from libraries unless we have a standard source of the routine at our disposal. Of course we are not trying to reinvent the wheel, as might easily be the case with algorithms that are numerically critical. However, we want a standard source in order to be able to transport a complete package.

We also try to raise some objections and see how they can be met:

- It is often said that optimal code and transportable code do not agree. Strictly spoken this is true. The important question is, however: how large is the difference in performance? Our idea is that often the difference is quite acceptable. It must be admitted that with the present compilers for supercomputers this rather depends on the problem at hand. The biggest problem is that there are cases where the compiler does not recognize our nicely vectorized program as being vectorized. Fortunately compilers are improving. Furthermore, as said above, we accept nonstandard versions if necessary, but only after a standard version has been written.
- The efficiency of an algorithm may depend on the computer architecture. If that is the case we introduce an implicit dependence on the computer installation. Still, in such a case we can try to code the algorithm in a transportable way. At least we will be able to run the program on another computer.
- Finally, there is evidence that the productivity of a programmer depends on the level of programming language (s)he uses, the productivity being larger for a higher language. The vector syntax may be considered as giving such a higher level language. Two things can be said. Firstly, as long as the vector syntax is not standardized we are minimizing our own time now, while maximizing other people's time (or our own time!) later. Secondly, in our experience the actual coding of an algorithm, once devised, is relatively easy, taking only moderate amounts of time.

4. ILLUSTRATIONS

In this section we give some illustrations of the principles mentioned in the previous section. As an example of the analysis of an algorithm we discuss the construction of the Fock matrix by numerical integration, as used in the HFS-LCAO method. As an example of the strategy to attain portability we consider a matrix multiplication routine.

4.1. Construction of Fock Matrix in HFS-LCAO

As mentioned in Section 2.2 one of the most time consuming steps in a Hartree-Fock-Slater-LCAO calculation is the construction of the Fock matrix in each cycle of the iterative procedure. For a general element of the Fock matrix we have

$$F_{\mu\nu} = \int dr\ \phi_\mu(r)\ f(r)\ \phi_\nu(r)$$

$$= \Sigma_k\ w(r_k)\ [\, f\phi_\mu(r_k)\phi_\nu(r_k) + \phi_\mu(r_k)\ f\phi_\nu(r_k)] , \qquad (4.1)$$

where the three-dimensional integration has been replaced by a sum over integration points k, the terms of which are weighted by a factor $w(r_k)$. Note that for convenience of presentation we simplified matters somewhat in Section 2.2. The complication arises because the Fock operator f in Eq. (4.1) is not simply a multiplicative term. In order to preserve hermiticity, we have to symmetrize the integrand.

As a first step in the analysis of the algorithm we note that the Fock operator may be separated in a local and a non-local term. This separation is useful because it is only the local part that changes in the iterative process. Thus, the calculation of the Fock matrix can be split in the calculation of the kinetic energy term T, which can be done once and for all, and the calculation of the potential, which has to be performed on each cycle of the iterative procedure:

$$F_{\mu\nu} = T_{\mu\nu} + V_{\mu\nu} \tag{4.2}$$

$$T_{\mu\nu} = \Sigma_k \, w(r_k) \, [\, t\phi_\mu(r_k) \; \phi_\nu(r_k) + \phi_\mu(r_k) \; t\phi_\nu(r_k)] \tag{4.3}$$

$$V_{\mu\nu} = \Sigma_k \, w(r_k) \; \phi_\mu(r_k) \; V(r_k) \; \phi_\nu(r_k) \tag{4.4}$$

This leads us to an observation of quite general nature. In the vectorization process the reconsideration of algorithms often leads to improvements which are also valid on a scalar processor. In many cases the deficiencies of the 'old' algorithm were already known (as it was the case for the HFS-LCAO program), but the improvement required too much effort to implement. Since the implementation of a vectorized algorithm often necessitates the rewriting of large parts of the program anyway, the additional effort of implementing the 'scalar' improvements may be small, making it an operation worth its time.

In Figure 3 we give the 'scalar' algorithm for the construction of the local part of the Fock matrix. We note that

```
      DO 40 K = 1,N
C
C        *** OMITTED HERE: CALCULATION OF U AT R(K) ***
C
         READ(ITAPE)  (PHI(MU),MU=1,M)
         DO 10 MU = 1,M
           TEMP(MU)  = PHI(MU) * U
   10    CONTINUE
C
         MUNU = 0
         DO 30 MU = 1,M
           DO 20 NU = 1,MU
             MUNU = MUNU + 1
             F(MUNU) = F(MUNU) + TEMP(MU) * PHI(NU)
   20      CONTINUE
   30    CONTINUE
C
   40 CONTINUE
```

Fig. 3. 'Scalar' algorithm for the construction of the local part of the Fock matrix.

this is not the actual code used in the HFS-LCAO program, but it is sufficiently close to characterize it. We have put the qualification scalar in quotes because no attempt has been made to vectorize the algorithm. It can be seen, however, that the algorithm contains vector operations, albeit of limited length. Furthermore, we note that although we have included a READ statement so as to remind us that the I/O needs to be done, we will not take this element into account when analysing the algorithm. For definiteness let us assume that the size of the basis is M (of order 10 to 100) and that the number of integration points is N (of order 1,000 to 10,000). It can be easily verified that the memory requirements of the algorithm are very modest, viz.

$$1/2\ M(M+5) + 1 \text{ words,} \tag{4.5}$$

assuming one word of storage for a real number. Further, we consider the efficiency η of the algorithm defined as the useful time (the time spent in the floating point operations) divided by the total time required (consisting of the useful time plus the startup time). Using the relation for the execution time of a vector operation, Eq. (3.1), it can be shown that

$$\eta = \frac{M(M+3)}{2(M+1)N_{1/2}+M(M+3)} \tag{4.6}$$

Hence we may characterize the 'scalar' algorithm as follows. The memory requirements are small. The effective vector length

```
C
C      *** OMITTED HERE: CALCULATION OF U(1..N) ***
C
       MUNU = 0
       DO 40 MU = 1,M
C
         READ(ITAPE)  (PHI(K,MU),K=1,N)
         DO 10 K = 1,N
           TEMP(K)  = PHI(K,MU)  * U(K)
   10    CONTINUE
C
         DO 30 NU = 1,MU
           MUNU = MUNU + 1
           DO 20 K = 1,N
           F(MUNU)  = F(MUNU)  + TEMP(K)  * PHI(K,NU)
   20      CONTINUE
   30    CONTINUE
C
   40 CONTINUE
```

Fig. 4. Vectorized algorithm for the construction of the local part of the Fock matrix.

is M/2, which is small compared to the startup length $N_{1/2}$ for vector operations on the Cyber 205. In addition, the ratio becomes even smaller when we use the molecular point group symmetry to construct a symmetry adapted basis, which causes the Fock matrix to become a direct sum of smaller Fock matrices.

As stated in Section 3.1 the most important point in vectorizing the algorithm is the identification of long vectors. Obviously this is very easy here: due to the nature of the problem the long vectors in the construction of the Fock matrix should pertain to the integration points, and not to the basis as they do in the 'scalar' algorithm. In Figure 4 we give the rewritten algorithm for the construction of the local part of the Fock matrix.

The analysis proceeds along the same lines as for the previous algorithm. For the memory requirements we find

$$(M+2)N + 1/2\ M(M+1) \text{ words,} \tag{4.7}$$

and for the efficiency

$$\eta = \frac{N}{N_{1/2}+N} \tag{4.8}$$

It can be seen, with the typical values for M and N in mind, that memory usage now becomes a problem. The reason is that the values of the basis functions need to be kept in storage during the construction process. In the scalar algorithm that did not bother us, it was merely a temporary vector. In the vectorized algorithm, however, the temporary vector has become a temporary matrix, and it does bother us. Of course, virtual memory allows us to get access to this amount of storage, but we will surely generate a tremendous number of page faults if the storage needed by the algorithm exceeds the available fast storage. Furthermore, the values of M and N given are typical for a scalar problem: they will increase when larger molecules are studied.

The efficiency of this algorithm is not a matter of concern now. The vector length will generally be so large that we will approach the asymptotic speed of the processor. Furthermore, the efficiency is independent of the size of the basis, which means that we obtain the full benefit of the use of point group symmetry when splitting the problem in sub-problems.

The characteristics of the algorithm are thus, that it is very efficient (when considering vector lengths only) but that the memory usage is the bottle-neck. On computers with virtual storage this may lead to paging problems. On computers without virtual storage it will strongly limit the size of problems that can be dealt with at all. Evidently a more flexible algorithm is needed.

In Figure 5 we give such an algorithm, which we coin the block vectorized algorithm. Its basic idea is that the vector length of the previous algorithm is normally so large, that we gladly sacrifice some of this length in order to reduce memory usage. The vectors over integration points are partitioned in

```
      DO 50 IBLOCK = 1,NBLOCK
C
C        *** OMITTED HERE: CALCULATION OF U(1..LBLOCK) ***
C
         MUNU = 0
         DO 40 MU = 1,M
C
            READ(ITAPE) (PHI(K,MU),K=1,LBLOCK)
            DO 10 K = 1,LBLOCK
               TEMP(K) = PHI(K,MU) * U(K)
   10       CONTINUE
C
            DO 30 NU = 1,MU
               MUNU = MUNU + 1
               DO 20 K = 1,LBLOCK
                  F(MUNU) = F(MUNU) + TEMP(K) * PHI(K,NU)
   20          CONTINUE
   30       CONTINUE
C
   40    CONTINUE
C
   50 CONTINUE
```

Fig. 5. Block vectorized algorithm for the construction of the local part of the Fock matrix.

blocks of length LBLOCK, the value of which may depend on the available storage. It may be noted that this partitioning of vectors is also found in algorithms for matrix multiplication when the matrices are too large to fit into core.

Table 1. Memory usage for algorithms for construction of local part of the Fock matrix (in 10^3 words). A refers to the scalar algorithm, B to the vectorized algorithm and C to the block vectorized algorithm with LBLOCK = 1000.

M	N	A	B	C
20	5000	0.25	110	21
20	10000	0.25	210	21
50	5000	1.38	260	51
50	10000	1.38	510	51

Table 2. Execution time for algorithms for construction of local part of the Fock matrix (in seconds). A refers to the scalar algorithm, B to the vectorized algorithm and C to the block vectorized algorithm with LBLOCK = 1000.

M	M	A	B	C	A/B	A/C
20	5000	0.297	0.024	0.027	12.5	11.2
20	10000	0.594	0.047	0.053	12.7	11.2
50	5000	0.810	0.137	0.154	5.9	5.3
50	10000	1.620	0.269	0.308	6.0	5.3

In Tables 1 and 2 we give some numerical data on storage and execution time of the three algorithms dealt with above. The numbers in Table 2 are the ones that can be calculated from the timing relation (3.1) with a realistic choice of parameters [12]. The execution time needed for the construction of the Fock matrix decreases by a factor of 5 to 10, depending on the molecular system investigated. It must be noted that this factor is attained by a change of algorithm alone. That is, we have not included the effect of using a vector processor instead of a scalar processor, or the improvement caused by a decreased clock cycle time. If these elements were included, much larger improvement factors would result.

We conclude this section with a remark on the data structure to be used by the program. The program sections treated above can be coded and tested separately on dummy data. However, in the real program the algorithm needs real data to act upon. The change of the algorithm also implies a change of the data structure used by the program. Since the program generates its own data, all the sections of the program that participate in calculating the data must be modified. It is this aspect of the vectorization of the HFS-LCAO program that requires most of the work involved.

4.2. Matrix Multiplication

As an example of implementation strategy we consider the matrix multiplication

$$C(NRA,NCB) := A(NRA,N) \times B(N,NCB) \ . \qquad (4.9)$$

The central operation in the calculation of C is

$$C(I,J) = C(I,J) + A(I,K) \times B(K,J) , \qquad (4.10)$$

embedded in three nested DO-loops: I = 1,NRA; J = 1,NCB; and K = 1,N. The number of floating point operations involved is $2 \times NRA \times N \times NCB$. Notice that this is a relatively large number, compared to, for instance, the number of floating point operations needed for the solution of a linear system of equations.

Identical results will be obtained for all 6 permutations of the 3 loops. Due to the different memory access patterns, however, the performance of the 6 algorithms will differ, even on a scalar computer. In Figure 6 we give a pictorial representation of the memory access patterns for each of the algorithms [10]. Vectors in this figure denote the elements accessed in the inner loop, while the middle loop corresponds to a loop over the vectors within a matrix.

The two algorithms which are most relevant for our purpose are the jki-algorithm (which has a linked triad as its inner loop) and the ijk-algorithm (which has a dot product as its inner loop). Algorithm jki is the usual solution. The important thing to notice is that in algorithm ijk the vector length used is NRA and in algorithm jki it is N. As long as these parameters are both large or of the same order, algorithm jki will be faster on the Cyber 205: it needs no gather and

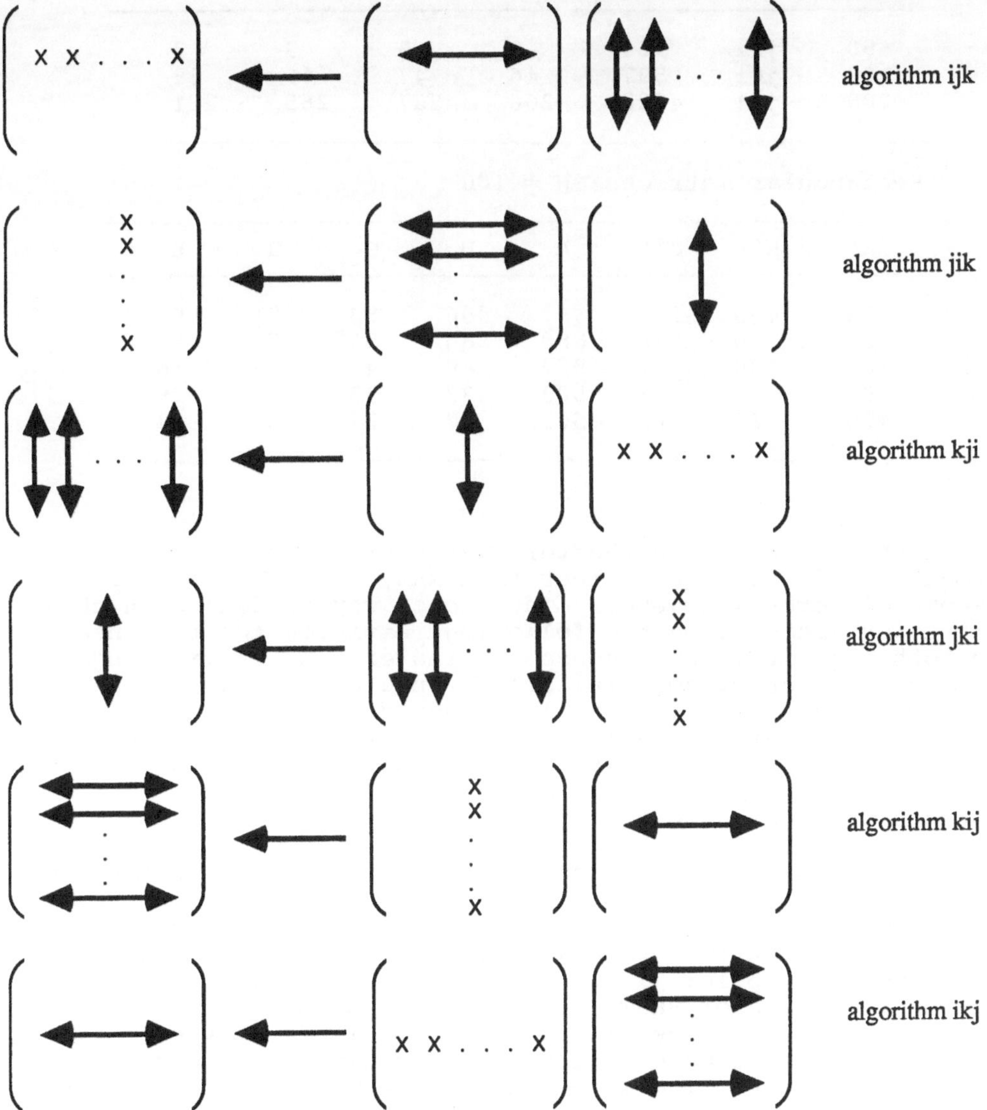

Fig. 6. Memory access patterns for the matrix multiplication C := AB [10].

Table 3. Execution time (in msec) of various versions of a matrix multiplication routine on the Cyber 205 at SARA (1 pipe). A-C refer to the same routine. A and B were translated with the FTN200 compiler: A - opt = DPRS; B - opt = DPRSV, unsafe; C was preprocessed by VAST, then compiled with opt = DPRS. D refers to a routine using the same algorithm in explicit vector syntax, compiled with opt = DPRS. E refers to the matrix multiplication routine QQMXMPY from QQLIB. The column "alg" gives the algorithm selected by the routine for A-D (see also text).

Square matrices: N = NRA = NCB

N	alg	A	B	C	D	E
50	jki	64	9	9	9	7
100	jki	507	46	47	46	38
200	jki	4028	266	267	265	231

Rectangular matrices: N = 100

NRA	NCB	alg	A	B	C	D	E
25	400	ijk	887	886	53	51	91
50	200	ijk	888	886	53	51	55
100	100	jki	507	46	47	46	38
200	50	jki	503	33	33	33	29
400	25	jki	501	27	27	27	24

the linked triad has a shorter startup-time than the dot product. If N becomes larger than NRA, the situation may become different, however. The proper way to deal with this situation seems to be the following: estimate the time needed by both algorithms and choose the faster one. Estimating the time needed can be done reliably by use of Eq. (3.1). We have written such a routine in standard Fortran 77. In order to keep the routine transportable we have defined the constants in Eq. (3.1) as a set of parameters, not only for this routine but for a whole package of routines [12]. They must be changed when going to another computer.

In Table 3 we give the execution time of the routine for matrices of various sizes and for various ways of generating the object program on the Cyber 205. Methods A-C start from the standard routine. A may be coined as scalar, B is vectorized by the compiler (level 640B was used), C is vectorized by the preprocessor VAST (level 1.11E), D is the hand-vectorized version, i.e. the same algorithm written in vector syntax. For comparison we have included the result of the matrix multiplication routine from the library QQLIB [11], written in machine language.

For square matrices our routine always selects algorithm jki: the linked triad has a smaller start-up time than the dot product; furthermore, algorithm jki does not need to gather rows of matrix A. From the results we see clearly that the vectorized code is much faster than the scalar code. For N = 200 a factor 15 is gained. Both ways of automatic vectorization (B,C) are just as good as hand vectorization (D). Only a small factor can be gained by going from Fortran to machine code (E). For rectangular matrices both algorithms may be selected. For N > NRA the larger overhead per vector operation can be compensated by their smaller number. Column B shows that the compiler does not succeed in vectorizing algorithm ijk (the trouble is the dot product). However, the preprocessor gives code (C) which is almost as good as the hand-vectorized code (D). Column E illustrates a well-known phenomenon: it is the algorithm that really matters. The routine from QQLIB does not take into account the dot product possibility and becomes inefficient for small values of NRA. This inefficiency becomes even more pronounced if smaller values for NRA are used.

CONCLUSIONS

The Hartree-Fock-Slater-LCAO method is very suitable for vectorization. Due to its numerical integration scheme and the avoidance of two-electron integrals its computational scheme can to a large extent be formulated in terms of dense matrices. Thus, the most time-consuming part of the program can be satisfactorily vectorized. In addition, the numerical integration scheme also allows large parts of the integral section to be vectorized.

The generation of vector instructions for operations on dense matrices can be performed by the compiler (sometimes in collaboration with a preprocessor). This allows one to use a strategy that stresses the portability of the program. We are of the opinion that the introduction of machine dependent elements, such as the use of special vector syntax, is an ill that should be postponed as long as possible. Good compilers help us in this respect.

With the advent of supercomputers many programs deserve a thorough rewrite. It is our experience with the vectorization of the HFS-LCAO program that it is not so much the vectorization of the time-consuming elements in the program, as the rewriting of those sections that deal with the modified data structures that consume most human time.

Finally, we note that the revision of algorithms for supercomputers often leads to improved performance on scalar computers as well, albeit not as drastic as the improvement of performance on the supercomputers. It is the latter improvement that makes it worth to invest time, and hence the former improvement may be considered as a useful spin-off. One should be careful to distinguish between the two, however.

ACKNOWLEDGMENTS

Useful discussions with E.J. Baerends, J.D. Schagen and J.G. Snijders are gratefully acknowledged. The vectorization of the Hartree-Fock-Slater-LCAO package is sponsored, in part, by the Stichting voor Fundamenteel Onderzoek der Materie (FOM) with financial support from the Nederlandse Organisatie voor Zuiver-Wetenschappelijk Onderzoek (ZWO).

REFERENCES

1. R. Ahlrichs, lecture, Utrecht (1985).
2. A. Szabo, and N.S. Ostlund, "Modern Quantum Chemistry", MacMillan, New York (1982).
3. E.J. Baerends, D.E. Ellis, and P. Ros, Chem. Phys., 2:41 (1973); E.J. Baerends, and P. Ros, ibid., 2:52 (1973); J.G. Snijders,and E.J. Baerends, Mol. Phys., 33:1651 (1977); T. Ziegler,and A. Rauk, Theor. Chim. Acta, 46: 1 (1977); J.G. Snijders,and E.J. Baerends, Mol. Phys., 36:1789 (1978); 38:1909 (1979).
4. W. Ravenek, and E.J. Baerends, J. Chem. Phys., 81:865 (1984); W. Ravenek,and F.M.M. Geurts, Chem. Phys., 90: 73 (1984); and references therein.
5. R.W. Hockney, and C.R. Jesshope, "Parallel Computers", Adam Hilger, Bristol (1981).
6. J.M. van Kats,and A. Rozendaal, "Input and output operations on a Cyber 205", Technical Report 20, ACCU, Utrecht, The Netherlands (1986).
7. A good introduction to the subject is: "Software Portability", P.J. Brown, ed., Cambridge University Press (1977).
8. F.P. Brooks, Jr., "The Mythical Man-month", Addison-Wesley, Reading, Massachusetts (1975).
9. Cyber 205 Service, "VAST Automatic Vectorizor User Guide", Control Data (1983).
10. J.J. Dongarra, F.G. Gustavson and A. Karp, SIAM Review, 26:91 (1984).
11. "QQLIB, a library of utility routines and mathematical algorithms on the Cyber 200", Control Data (1983).
12. W. Ravenek, "HFSLIB User Manual", Vrije Universiteit, Amsterdam, The Netherlands (1986).

Note added in proof

The present version of the FTN200 compiler at SARA recognizes the dot product for columns of two-dimensional arrays.

JACOBI-TYPE ALGORITHMS FOR EIGENVALUES ON VECTOR AND PARALLEL COMPUTERS°

M.H.C. Paardekooper

Tilburg University, The Netherlands

ABSTRACT

After a short introduction to Jacobi-like algorithms a review is given of a vector and a parallel implementation of the Jacobi method for symmetric matrices. In the last section a modification of Sameh's parallel eigen-algorithm is presented based on a problem formulation with so-called Euclidean parameters of non-orthogonal shears.

1. INTRODUCTION

The eigenvalue algorithm proposed by Jacobi, long before its time was ripe, was the favorite of numerical analists of yesterday because of its simplicity and reliability. The method of Jacobi (1846) has been rediscovered in 1950. The rise of the QR algorithm made an end to the revival of the Jacobi algorithm. But the growing importance of vectorized and parallel computing restored the interest in Jacobi-like algorithms (Brent and Luk, 1985; Luk, 1980; Modi and Pryce, 1985; Sameh, 1971; Stewart, 1985) and gave rise to its second revival.

An historical overview is given in Fig. 1. The centre column describes the highlights in the history of the Jacobi algorithm for symmetric matrices. The other columns give the overview for the skew symmetric case, $A = -A^T$, the normal case, $A A^* = A^* A$, the non-normal case and the singular value decomposition. The recent story of parallel eigenvalue computations starts in 1971 with Sameh's paper "Jacobi and Jacobi-like algorithms for a parallel computer" (Sameh, 1971).

It is the purpose of the present contribution to review some recent developments both theoretical and practical.

° This research is part of the VF-program "Parallele Algoritmiek", THD-WI-08/85-25 which has been approved by the Netherlands Ministry of Education and Sciences.

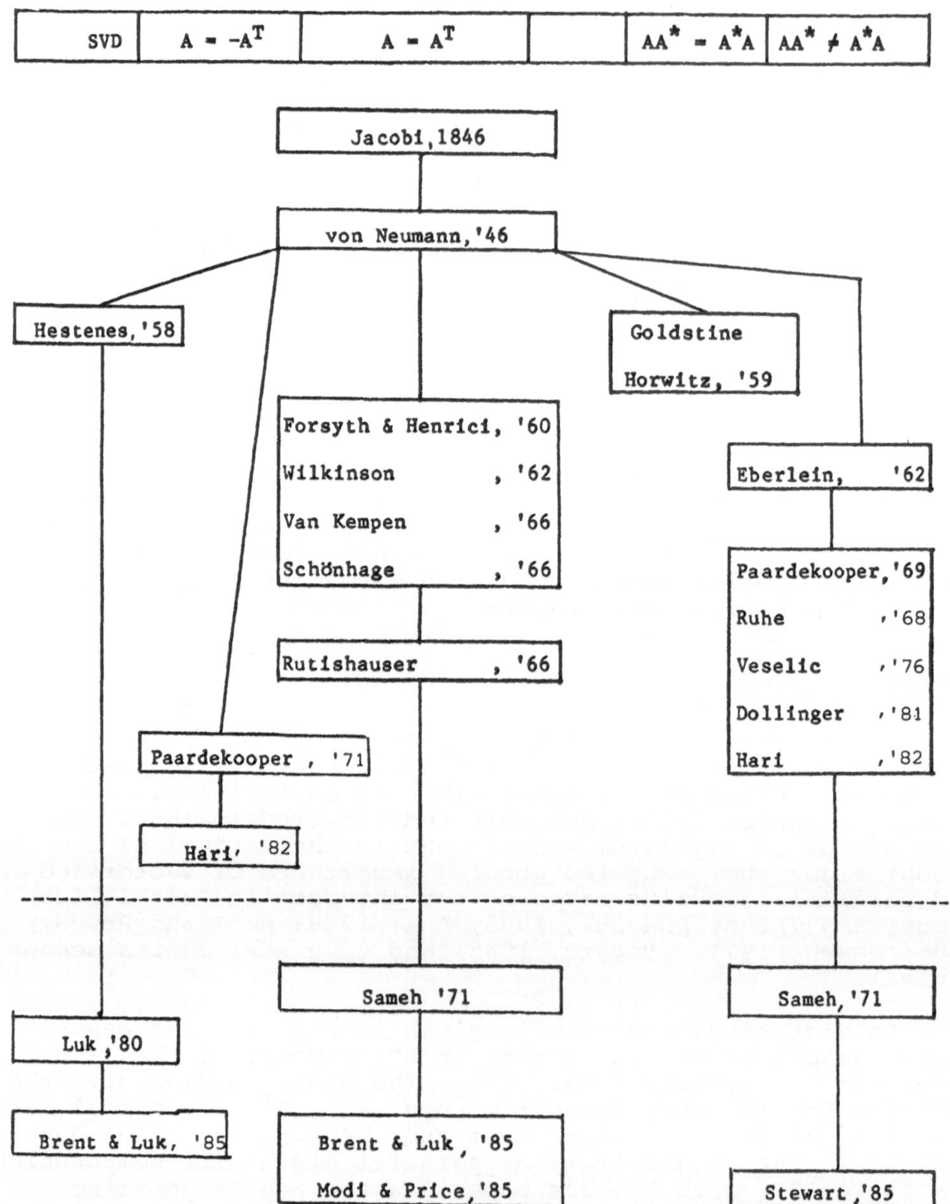

Fig. 1. Historical overview of Jacobi-like methods.

In Jacobi-like methods for the computation of the eigenvalues $\lambda_1,\ldots,\lambda_n$ in the spectrum $\sigma(A)$ of matrix $A \in \mathbb{R}^{n\times n}$ a sequence $\{A_k\}$, $A_0 := A$, is constructed in which the matrices $A_k = (a_{ij}^{(k)})$, $k = 0,1,\ldots$, are recursively defined by the relation

$$A_{k+1} = V_k^{-1}A_kV_k, \; k = 0,1,\ldots \; . \tag{1.1}$$

The matrix V_k differs from the unit matrix in the (ℓ_k,m_k)-plane. The non-trivial elements of V_k are the <u>Jacobi parameters</u>; they occur in the 2×2-matrix

$$\begin{pmatrix} v_{\ell_k,\ell_k}^{(k)} & v_{\ell_k,m_k}^{(k)} \\ v_{m_k,\ell_k}^{(k)} & v_{m_k,m_k}^{(k)} \end{pmatrix} = \begin{pmatrix} p_k & q_k \\ r_k & s_k \end{pmatrix} = \hat{V}_k \; . \tag{1.2}$$

This 2×2 matrix $\hat{V}_k$ will be called the (ℓ_k,m_k)-<u>restriction</u> of V_k. The choice of the successive pivot pairs (ℓ_k,m_k) is called the <u>pivot-strategy</u>. In several Jacobi-like processes the pivot-pairs are selected in some cyclic order.

In the Jacobi algorithm for the symmetric eigenvalue problem the Jacobi-parameters are

$$p_k = q_k = c_k = \cos\phi_k, \; r_k = -q_k = s_k = \sin\phi_k,$$

$$\phi_k \in (-\tfrac{\pi}{4},\tfrac{\pi}{4}] \; . \tag{1.3}$$

The rotation angle ϕ_k is chosen to minimize the sum of the squares of the non-diagonal elements:

$$\min_{-\frac{\pi}{4} < \phi_k \leq \frac{\pi}{4}} \sum_{i\neq j} (V_k^{-1}A_kV_k)_{ij}^2(\phi_k) \, .$$

In the non-normal case (Dollinger, 1981; Eberlein, 1962; Paardekooper,1969; Ruhe, 1968, Veselic, 1976) non-unitary shears are used in order to diminish, even to minimize, the Euclidean norms of the matrices in the sequence thus obtained. The minimization in each step of this process is more difficult than that in the symmetric case. After some easy but tedious calculation one observes that $\|V_k^{-1}V_kA_k\|_E$ is a rather simple function of the following variables:

$$x_k := |p_k|^2 + |q_k|^2, \; y := |r_k|^2 + |s_k|^2, \; z_k := p_k\bar{r}_k + q_k\bar{s}_k \tag{1.4}$$

to be called the <u>Euclidean parameters</u> of V_k.

Since the Euclidean norm of A is invariant under a unitary similarity transformation the optimal norm-reducing shear V_k is determined except for a unitary factor, shear Q_k. Matrices $S, P \in \mathbb{C}^{n\times n}$ will be called <u>row congruent</u> if $S = PU$ for some unitary matrix U. It is easy to see that S and P are row congruent iff $SS^* = PP^*$. Hence the shears V_k and W_k on the same pivot pair are row congruent iff they share their Euclidean parameters, for

$$\hat{V}_k\hat{V}_k^* = \begin{pmatrix} x_k & z_k \\ \bar{z}_k & y_k \end{pmatrix} . \tag{1.5}$$

This property of row contingent shears explains that $\|V_k^{-1}A_kV_k\|_E$ is a function of x_k, y_k and z_k and this quality will be used intensively in section 3.

Section 2 is concerned with the Jacobi process for symmetric matrices. Sequential as well as parallel algorithms are described. We emphasize the nice parallel Jacobi-algorithm of Brent-Luk (1985) for a systolic array computer. The complexity result of their algorithm is impressive.

Section 3 reviews some Jacobi-like eigen algonrithms for non-normal matrices. The description of Eberlein's algorithm (Eberlein, 1962; Paardekooper, 1969) with Euclidean parameters introduces our modification of Sameh's parallel algorithm (Sameh, 1971).

2. JACOBI METHODS FOR THE SYMMETRIC EIGENVALUE PROBLEM

2.1. In each step of the Jacobi method for the symmetric eigenproblem the norm of the non-diagonal part of the new matrix $A^{(k+1)} = V_k^T A^{(k)} V_k$ is minimized with respect to the rotation parameter $\phi_k \in (-\frac{\pi}{4},\frac{\pi}{4}]$ of the Jacobi parameters of V_k. The minimum of

$$f(A^{(k+1)}) = \sum_{i\neq j} (V_k^T A^{(k)} V_k)_{ij}^2$$

is obtained if $A^{(k+1)}_{\ell_k,m_k} = 0$. This annihilation of the (ℓ_k,m_k)-th element occurs iff

$$\tan 2\phi_k = 2\,\mu_k/\nu_k , \tag{2.1}$$

where

$$\mu_k = a^{(k)}_{\ell_k,m_k} , \quad \nu_k = a^{(k)}_{\ell_k,\ell_k} - a^{(k)}_{m_k,m_k} .$$

As an easy consequence of the orthogonality of V_k one finds

$$f(A^{(k+1)}) - f(A^{(k)}) = 2\mu_k^2 . \tag{2.2}$$

In the classical Jacobi iteration μ_k is the largest off-diagonal element in modulus. This optimal pivot-strategy is a special case of that described in:

Theorem 2.1. If $\mu_k^2 \geqq \frac{1}{n(n-1)} f(A^{(k)})$ for each $k \in \mathbb{N}$, then

$$\lim_{k\to\infty} A^{(k)} = \mathrm{diag}(\lambda_j)$$

Proof. With (2.2) we have

$$f(A^{(k+1)}) = f(A^{(k)}) - 2\mu_k^2 \leqq (1 - \frac{2}{n(n-1)}) f(A^{(k)}), \ k \in \mathbb{N}.$$

Hence $f(A^{(k)}) \to 0(,k\to\infty)$. Let be $\delta = \min\{|\lambda_i-\lambda_j|;\lambda_i,\lambda_j \in \sigma(A), \lambda_i \neq \lambda_j\}$. Then $f(A^{(k)}) \leqq \frac{1}{16}\delta^2$ for each k larger than some N. For $k > N$ there exists a permutation π of $\{1,\dots,n\}$ such that $(\lambda_{\pi(i)}-a_{ii}^{(k)})^2 \leqq f(A^{(k)}) \leqq \frac{1}{16}\delta^2$ for each $i \in \{1,\dots,n\}$, as follows from the Hoffmann-Wieland theorem. Since $|\phi_k| \leqq \frac{\pi}{4}$ and $|a_{\ell_k,\ell_k}^{(k+1)} - a_{\ell_k,\ell_k}^{(k)}| = |a_{m_k,m_k}^{(k+1)} - a_{m_k}^{(k)}| = |\mu_k \tan\phi_k| \leqq \frac{1}{4}\delta$, $a_{\ell_k,\ell_k}^{(k+1)} \in [\lambda_{\pi(\ell_k)} - \delta/4, \lambda_{\pi(\ell_k)} + \delta/4]$. Similarly a_{m_k,m_k}^{k+1} remains in the corresponding interval around $\lambda_{\pi(m_k)}$. The stationary matching of diagonal elements and eigenvalues together with $(\lambda_{\pi(i)}-a_{ii}^{(k)})^2 \leqq f(A^{(k)}) \to 0(,k\to\infty)$ implies the theorem.□

As an alternative for the pivot strategy mentioned in theorem 2.1 a cyclic method can be used, especially the row serial method. Then the elements are annihilated in the cyclic order (1,2),(1,3),...,(1,n),(2,3),...,(n-1,n),(1,2),... . Forsyth and Henrici (1960) proved the convergence of the serial method. The ultimate convergence is quadratic, i.e. $f(A^{k+M}) \leq \text{constant}.f^2(A^k)$, k large enough, where $M = 1/2\, n(n-1)$; this has been investigated in (van Kempen, 1966; Schönhage, 1964; Wilkinson, 1962).

2.2. The transformation $A^{(k+1)} = V_k^T A^{(k)} V_k$ requires 4n multiplications plus 2n additions. On a supercomputer like the CYBER 205 it is recommendable to speed up this time-consuming process with linked triads (Paardekooper, 1986). Let $A^{(k)} = D_k H^{(k)} D_k$, where $D_k = \mathrm{diag}(d_1^{(k)},\dots,d_n^{(k)})$, with $D_0 = I$. Since

$$A^{(k+1)} = D_{k+1}\, H^{(k+1)}\, D_{k+1} = V_k^T\, D_k H^{(k)} D_k V_k, \tag{2.3}$$

where

$$\hat{V}_k = \begin{pmatrix} c_k & s_k \\ -s_k & c_k \end{pmatrix}$$

the obvious updating for D_k:

$$d_i^{(k+1)} = \begin{cases} d_i^{(k)} & , i \neq \ell_k, m_k \\ c_k d_i^{(k)} & , i \in \{\ell_k, m_k\} \end{cases} \tag{2.4}$$

brings about linked triads in the updating of $H^{(k)}$, viz.

$$H^{(k+1)} = J_k^T\, H^{(k)}\, J_k . \tag{2.5}$$

The (ℓ_k,m_k)-restriction of J_k is

$$\begin{pmatrix} 1 & \beta_k \\ -\alpha_k & 1 \end{pmatrix}, \begin{cases} \beta_k = t_k\, d_{m_k}^{(k)}/d_{\ell_k}^{(k)} \\ \alpha_k = t_k\, d_{\ell_k}^{(k)}/d_{\ell_k}^{(k)} \end{cases}, \quad t_k = \tan\phi_k.$$

Let v_k be the ℓ_k-th column of $H^{(k)}$ and w_k the m_k-th column of $H^{(k)}$. The corresponding columns of $H^{(k)}J_k$ are $\tilde{v}_k = v_k - \alpha_k w_k$, $w_k' = w_k + \beta_k v_k = (1+\alpha_k\beta_k)(w_k+\tilde{\beta}_k\tilde{v}_k)$ with $\tilde{\beta}_k = \beta_k/(1+\alpha_k\beta_k)$. With the updating scheme

$$\begin{cases} \tilde{v}_k = v_k - \alpha_k \, w_k \\ w_k' = w_k + \tilde{\beta}_k \, \tilde{v}_k \end{cases} , \quad \begin{cases} d_{\ell_k}^{(k+1)} = c_k \, d_{\ell_k}^{(k)} \\ d_{m_k}^{(k+1)} = c_k^{-1} \, d_{m_k}^{(k)} \end{cases} , \qquad (2.6)$$

one avoids the necessity to copy v_k. In a similar way the new rows are computed. The multiplication $D_{k+1}H^{(k+1)}D_{k+1}$ is postponed until the end of the process. We conclude that variant (2.6) of a serial Jacobi-method is appropriate for a CYBER-like supercomputer (Paardekoper, 1986).

2.3. The Jacobi methods described so far all were sequential. In essence the Jacobi-method with its nested loop for which computations are almost identical over the entire index set $\{(i,j)\}$ is pre-eminently suited for processing on an array processor with regular dataflow (Brent and Luk, 1985; Luk, 1980; Modi and Pryce, 1985). The systolic implementation of the Jacobi method of Brent and Luk (1985) has a high degree of modularity, absence of long data paths, near-by connectivity and a simple synchronizing mechanism.

Assume n to be even. Consider the parallel Jacobi-like updating of the column pairs $2\ell - 1$, $2\ell(,\ell=1,\ldots,n/2)$ and thereafter the corresponding updating of the same pairs of rows. This achieves the annihilation of the elements $a_{2\ell-1,2\ell}$. In order to achieve an analogous updating of another set of n/2 column-row pairs consider the permutation π of $\{1,\ldots,n\}$ such that

$$\pi(i) = \begin{cases} i & , \quad i = 1 \\ i+2 & , \quad i = 2,4,\ldots,n-2 \\ i-2 & , \quad i = 5,7,\ldots,n-1 \\ i-1 & , \quad i = 3,n \end{cases} .$$

The repeated execution of this caterpillar permutation is illustrated in figure 2.

The repeated annihilation of codiagonal elements can be performed on a systolic array processor, see figure 3. The square n/2 by n/2 array consists of processors P_{ij}, $i,j = 1,\ldots,n/2$, each containing the corresponding 2×2 submatrix,

$$\begin{array}{ccccccccc} 1 & & 3 & \longrightarrow & 5 & \longrightarrow & \cdots & \longrightarrow n-3 & \longrightarrow n-1 \\ & \nearrow & & & & & & & \downarrow \\ 2 & \longleftarrow & 4 & \longleftarrow & 6 & \longleftarrow & \cdots & \longleftarrow n-2 & \longleftarrow n \end{array}$$

Fig. 2. Caterpillar permutation $\pi(1,\ldots,n) = (1,4,2,6,\ldots,n-3,n-1)$.

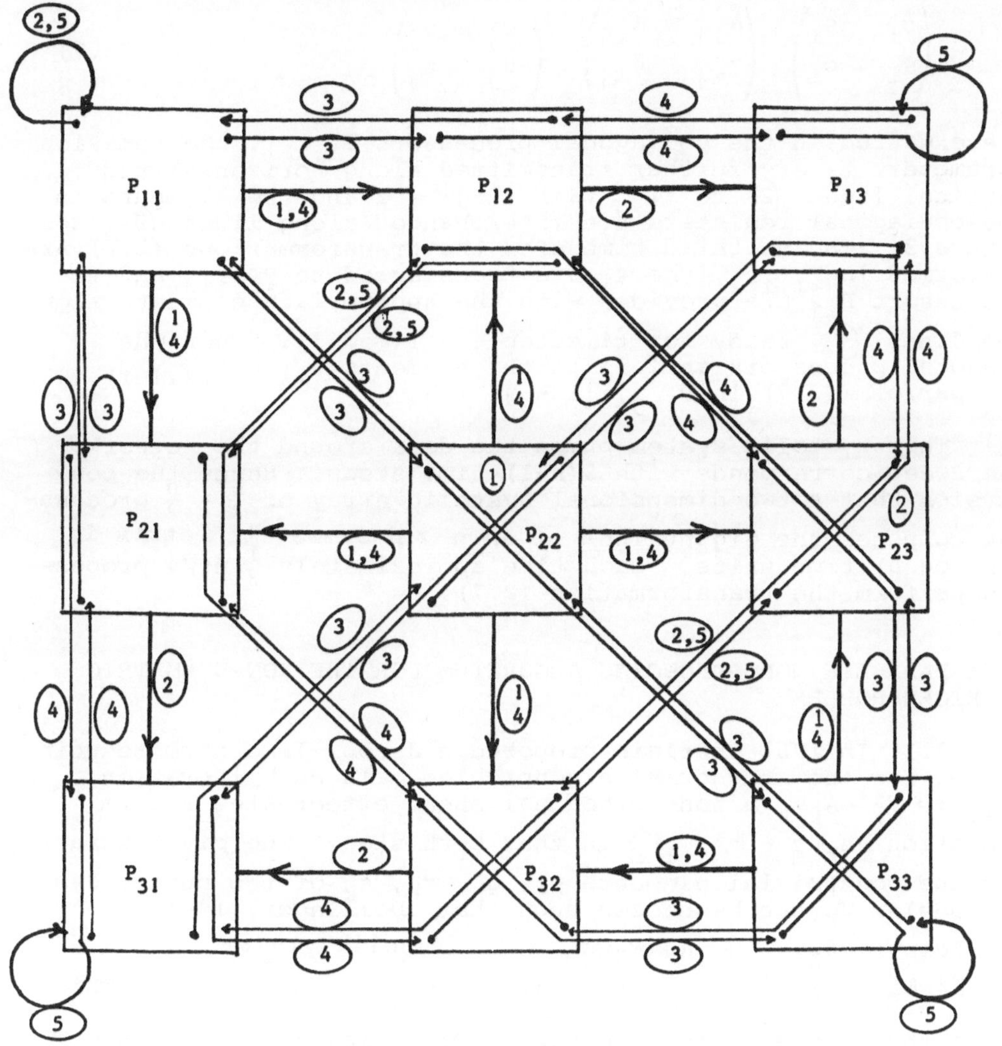

Fig. 3. Systolic array. The number of the lines indicate the time they are active.

viz.

$$\tilde{A}_{ij} = \begin{pmatrix} \alpha_{ij} & \beta_{ij} \\ \gamma_{ij} & \delta_{ij} \end{pmatrix} = \begin{pmatrix} a_{2i-1,2j-1} & a_{2i-1,2j} \\ a_{2i,2j-1} & a_{2i,2j} \end{pmatrix}, \quad i,j = 1,\ldots,n/2.$$

In the first time step the processors P_{ii} compute $\begin{pmatrix} c_i & -s_i \\ s_i & c_i \end{pmatrix}$ and annihilate the elements β_{ii} $(=\gamma_{ii})$. Horizontal and vertical output lines ① transport the rotation parameters $t_i = s_i/c_i$ away from the diagonal. In the second time step the transformations

$$\begin{pmatrix} c_i & -s_i \\ s_i & c_i \end{pmatrix} \begin{pmatrix} \alpha_{ij} & \beta_{ij} \\ \gamma_{ij} & \delta_{ij} \end{pmatrix} \begin{pmatrix} c_j & s_j \\ -s_j & c_j \end{pmatrix} \tag{2.7}$$

are executed in the codiagonal processors $P_{i,i\pm 1}$, the rotation parameters t_i are further transmitted along horizontal and vertical lines ② to $P_{i,j}$ with $|i-j| = 2$ and the elements in the codiagonal registers are interchanged along lines ②, see figure 3. In the third time step the transformations (2.7) are performed in $P_{i,i\pm 2}$, the t_i are transferred to $P_{i,i\pm 3}$ and the processors P_{ij} are provided with the appropriate elements along the lines ③, ready for timestep 4 = 1(mod 3). Then the diagonal processors annihilate the elements $\beta_{ii} = a_{\pi(2i-1),\pi(2i)}$ and perform (2.7) in $P_{i,j}$, $|i - j| = 3$.

This systolic system pumps the data around the network. One sweep corresponds with 3(n-1) time steps. Hence the conclusion that a two-dimensional systolic array of $\frac{n}{2} \times \frac{n}{2}$ procession computes the eigenvalues of an $n \times n$ symmetric matrix in $O(n \log n)$ time units. Each time approximately $\frac{1}{3}\, n^2/4$ processors perform the transformation (2.7).

3. A PARALLEL NORM-REDUCING ALGORITHM FOR THE NON-SYMMETRIC EIGENPROBLEM

3.1. In 1962 Eberlein proposed a Jacobi-like norm-reducing method for the non-normal eigenproblem. In each iteration $A_{k+1} := V_k^{-1} A_k V_k$ a non-orthogonal shear effectuates a norm reduction $\|A_k\|_E^2 - \|A_{k+1}\|_E^2$ in that k-th step. The pivot strategy and the non-trivial elements p_k, q_k, r_k, s_k of the successive unimodular V_k can be chosen such that (Dollinger, 1981; Eberlein, 1962; Paardekooper, 1969; Veselić, 1976) $\lim_{k\to\infty} \|A_k\|_E^2 = \sum_{j=1}^{n} |\lambda_j|^2$, $\lambda_j \in \sigma(A)$, eigenvalue of $A = A_0$. This means that sequence $\{A_k\}$ converges to normality (Elsner and Paardekooper; Paardekooper, 1969). The unimodularity of V_k implies that the Euclidean parameters (x_k, y_k, z_k) of the real shear V_k satisfy the conditions

$$x_k,\ y_k > 0, \quad x_k y_k - z_k^2 = 1 . \tag{3.1}$$

Easy calculations give the result already announced in the introduction.

Theorem 3.1 (Paardekooper, 1969). If V is a real unimodular shear with pivot pair (ℓ,m) and Euclidean parameters $(x,y,z) \in \mathbb{R}^3$, then

$$\|V^{-1}AV\|_E^2 = f(x,y,z) + \sigma + e , \tag{3.2}$$

with

$$\sigma = \sum_{i,j\notin\{\ell,m\}} a_{ij}^2,\quad e = (a_{\ell\ell}+a_{mm})^2 - 2(a_{\ell\ell}a_{mm}-a_{\ell m}a_{m\ell}) \text{ and}$$

$$f(x,y,z) = \alpha x + \beta y + 2\gamma z + (-\lambda x+\mu y+\nu z)^2 \tag{3.3}$$

where

$$\alpha = \sum_{i\neq\ell,m} (a_{i\ell}^2+a_{mi}^2),\quad \beta = \sum_{i\neq\ell,m} (a_{\ell i}^2+a_{im}^2),$$

$$\gamma = \sum_{i\neq\ell,m} (a_{i\ell}a_{im} - a_{\ell i}a_{mi})$$

and

$$\lambda = a_{m\ell},\ \mu = a_{\ell m},\ \nu = a_{\ell\ell} - a_{mm}.\square$$

So the minimization of f on $\mathcal{H} := \{(x,y,z)\,|\,x,y > 0,\ xy - z^2 = 1\}$ provides the optimal norm-reducing unimodular shears. An accurate analysis of that minimization problem leads to

Theorem 3.2 (Paardekooper, 1969). Let be $D = \alpha\mu - \beta\lambda - \gamma\nu$, $E = \nu^2+4\lambda\mu$, $F = \alpha\beta - \gamma^2$. If D and F are both not equal to zero then f is minimal on H in the point

$$x = \frac{2\mu D - \beta(\rho-E)}{\rho(\rho-E)},\quad y = \frac{-2\mu D - \beta(\rho-E)}{\rho(\rho-E)},\quad z = \frac{-\mu D + \beta(\rho-E)}{\rho(\rho-E)}, \tag{3.4}$$

where ρ is the unique root of the quadratic equation

$$(\rho-E)^2(\rho-F) + D^2(2\rho-E) = 0$$

for which holds $\rho < \min\{0,E\}$. The infimum of f on $\mathcal{H}$ is not assumed when $D = 0 \wedge F = 0 \wedge(\alpha+\beta\neq 0 \vee(E=0 \wedge \lambda\neq\mu))$. Then the interaction of the planes $\alpha x + \beta y + 2\gamma z = 0$ and $-\lambda x + \mu y + \nu z = 0$ is in the tangent cone $xy - z^2 = 0$ of $\mathcal{H}$.$\square$

With the new variables

$$w := (x-y)/2,\quad t := t(w,z) = (x+y)/2 = (1+w^2+z^2)^{1/2} \tag{3.5}$$

we get $\|V^{-1}AV\|_E^2 = g(w,z) + \sigma + e$, where

$$g(w,z) = (\alpha+\beta)t + (\alpha-\beta)w + 2\gamma z + ((\mu-\lambda)t - (\mu+\lambda)w + \nu z)^2. \tag{3.6}$$

Now let be $C = C(A) = A^TA - AA^T$, the <u>commutator</u> of A, being a measure of non-normality. One easily finds

Theorem 3.3 (Paardekooper, 1969). Grad $g(0,0) = (c_{mm}-c_{\ell\ell}, 2c_{\ell m})^T$. Moreover $\|\tilde{V}^{-1}A\tilde{V}\|_E = \min\{\|V^{-1}AV\|_E\,|\,V$ unimodular (ℓ,m)-shear$\}$ Iff $\tilde{c}_{mm} - \tilde{c}_{\ell\ell} = \tilde{c}_{\ell m} = 0$, where $(\tilde{c}_{ij}) = C(\tilde{V}^{-1}A\tilde{V})$, the commutator of $\tilde{V}^{-1}A\tilde{V}$.$\square$

This result gives an indication for an appropriate pivot strategy: choose in each step (ℓ,m) such that $\|\text{grad } g(0,0)\|^2 = (c_{mm}-c_{\ell\ell})^2 + 4c_{\ell m}^2$ is maximal. These choices of (ℓ_k,m_k) together with optimal norm-reducing shears V_k guarantee that $C(A_k) \to 0\ (,k\to\infty)$ (Dollinger, 1981; Eberlein, 1962; Paardekooper, 1969; Veselić, 1976).

3.2. The purpose of this subsection is to present an improved version of Sameh's parallel norm-reducing process (Sameh, 1971). Therefore we assume A to be of even order $n = 2k$ and partitioned as follows

$$A = \begin{pmatrix} A_{11} & A_{12} & \cdots & A_{1k} \\ A_{21} & A_{22} & \cdots & A_{2k} \\ \vdots & \vdots & & \\ A_{k1} & A_{k2} & \cdots & A_{kk} \end{pmatrix}, \tag{3.7}$$

where each submatrix $A_{\ell m}$ is given by

$$A_{\ell m} = \begin{pmatrix} a_{2\ell-1,2m-1} & a_{2\ell-1,2m} \\ a_{2\ell,2m-1} & a_{2\ell,2m} \end{pmatrix}, \quad \ell, m = 1,\dots,k\,. \tag{3.8}$$

For convenience we define

$$\lambda_{\ell m} := a_{2\ell,2m-1},\ \mu_{\ell m} := a_{2\ell-1,2m},\ \nu_{\ell m} := a_{2\ell-1,2\ell-1} - a_{2\ell,2\ell}.$$

Let be

$$\tilde{A} = V^{-1}AV \tag{3.9}$$

where $V = \operatorname{diag}(S_1, S_2, \dots, S_k)$ with

$$S_i = S = \begin{vmatrix} p & q \\ r & s \end{vmatrix}, \quad ps - qr = 1,\ i = 1,\dots,k.. \tag{3.10}$$

The computation of $V^{-1}AV$ is readily adapted to parallel computation: firstly simultaneous updating of the k column pairs, next the simultaneous updating of the k rows: $V^{-1}(AV)$. With Euclidean parameters (x,y,z) of S in (3.10) one obtains, analogous to theorem 3.1,

Theorem 3.4. If V is $\operatorname{diag}(S_1,\dots,S_k)$ where S_i as given in (3.10) then $\|V^{-1}AV\|_E^2 = h(x,y,z)+K$, where

$$h(x,y,z) = \sum_{\ell,m=1}^{k} (-\lambda_{\ell m}x + \mu_{\ell m}y + \nu_{\ell m}z)^2 \tag{3.11}$$

and

$$K = \sum_{\ell,m=1}^{k} (\operatorname{tr}^2 A_{\ell m} - 2\det(A_{\ell m})).\square$$

The minimization of h on $\mathcal{H}$ leads to a generalized eigenproblem: $\det(B^TB-\rho H) = 0$, where

$$B = \begin{pmatrix} b_1, b_2, b_3 \end{pmatrix} \in \mathbb{R}^{\frac{1}{4}n^2\times 3}, \quad H = \begin{pmatrix} 0 & \frac{1}{2} & 0 \\ \frac{1}{2} & 0 & 0 \\ 0 & 0 & -1 \end{pmatrix}, \text{ with}$$

$$b_1 = -(\lambda_{11},\lambda_{12},\dots,\lambda_{1k},\lambda_{21},\dots,\lambda_{kk})^T$$

$$b_2 = (\mu_{11},\mu_{12},\dots,\mu_{1k},\mu_{21},\dots,\mu_{kk})^T \qquad (3.12)$$

$$b_3 = (\nu_{11},\nu_{12},\dots,\nu_{1k},\nu_{21},\dots,\nu_{kk})^T$$

With the usual compactness, continuity and convexity arguments one reaches
Theorem 3.5. If rank (B) = 3, B = QR with R upper triangular then the vector of optimal Euclidean parameters of S in V is eigenvector corresponding with the unique positive eigenvalue of $R^{-T}H\,R^{-1}$.□

In the case of collinearity in the matrix B it may occur that the function h is not minimal on $\mathcal{H}$:
Theorem 3.6. If rank (B) = 1 and $\mathrm{im}(B^T) = \{\tau(p_1,p_2,p_3)^T \mid \tau \in \mathbb{R}\}$ then

i) $\min \{h(x,y,z) \mid x,y > 0,\ xy - z^2 = 1\} > 0$ if $p_3^2-4p_1p_2 < 0$.

ii) $\inf \{h(x,y,z) \mid x,y > 0,\ xy - z^2 = 1\} = 0$ if $p_3^2-4p_1p_2 \geqq 0$; this infinum is assumed iff $p_3^2-4p_1p_2 > 0$. (3.13)

If rank (B) = 2 and $\ker(B) = \{\tau(p_1,p_2,p_3)^T \mid \tau \in \mathbb{R}\}$, then

i) $\min \{h(x,y,z) \mid x,y > 0,\ xy - z^2 = 1\} > 0$ if $p_1p_2 < p_3^2$.

ii) $\inf \{h(x,y,z) \mid x,y > 0,\ xy - z^2 = 1\} = 0$ if $p_1p_2 \geqq p_3^2$; this infinum is assumed iff $p_1p_2 > p_3^2$.□ (3.14)

The resemblance of the transforms (3.10) and the Eberlein shear transform is also manifest in the analogue of theorem 3.3. Let be $w = (x-y)/2$, $t = (x+y)/2$ as in (3.5). Then

$$\|V^{-1}AV\| = \|(b_1+b_2)t + (b_1-b_2)w + b_3z\|^2 + K =: g(w,z;A) + K \qquad (3.15)$$

With these new variables w, z and $t = t(w,z) = (1+z^2+w^2)^{1/2}$ we get the analogue of theorem 3.3 by simple calculations:
Theorem 3.7. $\mathrm{Grad}\, g(0,0;A) = \sum_{\ell=1}^{k} (c_{2\ell-1,2\ell-1}-c_{2\ell,2\ell}, 2c_{2\ell-1,2\ell})^T$. Moreover the parallel identical shear transformation $V^{-1}AV$ gives an optimal norm-reduction iff $\sum_{\ell=1}^{k} (\tilde{c}_{2\ell-1,2\ell-1}-\tilde{c}_{2\ell,2\ell}, 2\tilde{c}_{2\ell-1,2\ell}) = (0,0)$, where $(\tilde{c}_{ij}) = C(V^{-1}AV)$.□

3.3. Since $\mathrm{grad}(0,0;A) = \sum_{\ell=1}^{k} (c_{2\ell-1,2\ell-1}-c_{2\ell 2\ell}, 2c_{2\ell-1,2\ell})$, a prologue transformation (Sameh, 1971) with a well chosen direct sum $Q = \mathrm{diag}(Q_1,Q_2,\dots,Q_k)$ of orthogonal shears

$Q_\ell, \ell=1,\dots,k$, may enlarge the gradient. Let be $(c'_{ij}) = C(Q^TAQ)$ then

$$v'_\ell := \begin{pmatrix} c'_{2\ell-1,2\ell-1} - c'_{2\ell,2\ell} \\ 2c'_{2\ell-1,2\ell} \end{pmatrix} = Q_\ell^{-2} \begin{pmatrix} c_{2\ell-1,2\ell-1} - c_{2\ell,2\ell} \\ 2c_{2\ell-1,2\ell} \end{pmatrix} =: Q_\ell^{-2} v_\ell \tag{3.16}$$

Each v'_ℓ has the same direction by an appropriate choice of the Q_ℓ. Then

$$\|g(0,0;Q^TAQ)\| = \|\sum_{\ell=1}^{k} v'_\ell\| = \sum_{\ell=1}^{k} \|v'_\ell\| = \sum_{\ell=1}^{k} \|v_\ell\| \tag{3.17}$$

The vectors $v_\ell \in \mathbb{R}^2$ will be rectified with simultaneous Jacobi annihilations applied to C such that

$$c'_{2\ell-1,2\ell} = 0,\ c'_{2\ell-1,2\ell-1} - c'_{2\ell,2\ell} = \|v_\ell\|,\ \ell = 1,\dots,k. \tag{3.18}$$

The preconditioning $A \to A' = Q^TAQ$ simplifies the performance of the first steepest descent iteration for the minimization of g.

A lower bound of $\|A\|_E^2 - \|V^{-1} Q^TAQV\|$ is given in

Theorem 3.8. Let be $Q = \mathrm{diag}(Q_1,\dots,Q_k)$ a direct sum of orthogonal shears such that $(c'_{ij}) = C(Q^TAQ)$ satisfies (3.18). Then there exists a diagonal matrix $V = \mathrm{diag}(x,x^{-1},x,\dots,x^{-1})$ such that

$$\|A\|_E^2 - \|V^{-1} Q^TAQV\|_E^2 \geq \frac{1}{8} \|A\|_E^{-2} \sum_{\ell=1}^{k} \|v_\ell\|^2 . \tag{3.19}$$

Proof. Let be $A' = Q^TAQ$ and $B' = \begin{pmatrix} b'_1 & b'_2 & b'_3 \end{pmatrix}$ similar to (3.12). Then $\|V^{-1}A'V\|_E^2 := \|x^2b'_1 + x^{-2}b'_2\|^2$ + constant. The minimizing x gives

$$\|A\|_E^2 - \|V^{-1} Q^TAQV\|_E^2 = (\|b'_1\| - \|b'_2\|)^2 .$$

Since, assuming $b'_1, b'_2 \neq 0$,

$$\|b'_1\| - \|b'_2\| = (\|b'_1\| + \|b'_2\|)^{-1}(\|b'_1\|^2 - \|b'_2\|^2) = \frac{1}{2}(\|b'_1\| + \|b'_2\|^{-1} g_w(0,0;A'))$$

we obtain from $\|b'_1\| + \|b'_2\| \leq \|A\|_E \sqrt{2}$ and

$$g_w(0,0;A') = \sum_{\ell=1}^{k} (c'_{2\ell-1,2\ell-1} - c'_{2\ell,2\ell}) = \sum_{\ell=1}^{k} \|v_\ell\|$$

that

$$\|A\|_E^2 - \|V^{-1}A'V\|_E^2 \geq \frac{1}{8} \|A\|_E^{-2} (\sum_{\ell=1}^{k} \|v_\ell\|)^2 \geq \frac{1}{8} \|A\|_E^{-2} \sum_{\ell=1}^{k} \|v_\ell\|^2 .$$

In the case $b'_2 = 0$ and $b'_1 \neq 0$ the same bound holds: take

$x^4 < 1 - \frac{1}{8}\|A\|_E^{-2}\|b_1'\|^2$.□

For the main theorem 3.10 finally we need a modification of a lemma in (Eberlein, 1962).

Theorem 3.9. There exists a set of k disjunct index pairs (ℓ_j, m_j) with $\ell_j \neq m_j$, $j = 1,\ldots,k$ such that

$$\sum_{j=1}^{k} ((c_{\ell_j,\ell_j} - c_{m_j,m_j})^2 + 4c^2_{\ell_j,m_j}) \geqq \frac{4}{n-1}\|C(A)\|_E^2 . \qquad (3.20)$$

Proof. Since $\sum_{\ell=1}^{n} c_{\ell\ell} = 0$, $\sum_{\ell\neq m} (c_{\ell\ell}-c_{mm})^2 = (2n-1)\sum_{\ell=1}^{n} c_{\ell\ell} - \sum_{\ell\neq m} c_{\ell\ell}c_{mm}$ and $0 = (\sum_{\ell=1}^{n} c_{\ell\ell})^2 = \sum_{\ell=1}^{n} c^2_{\ell\ell} + \sum_{\ell\neq m} c_{\ell\ell}c_{mm}$, we have for $n \geqq 2$: $\sum_{\ell\neq m} (c_{\ell\ell}-c_{mm})^2 = 2n\sum_{\ell=1}^{n} c^2_{\ell\ell} \geqq 4\sum_{\ell=1}^{n} c^2_{\ell\ell}$. Consequently

$$\sum_{\ell\neq m} ((c_{\ell\ell}-\ell_{mm})^2 + 4c^2_{\ell m}) \geqq 4\,\|C(A)\|^2 .$$

Hence the mean of $\sum_{j=1}^{k} ((c_{\ell_j,\ell_j} - c_{m_j,m_j})^2 + 4c^2_{\ell_j,m_j})$ over the sets ω of k distinct index pairs (ℓ_j,m_j) satisfies (3.20) for there are $n!/(k!2^k)$ such sets ω and each pair (ℓ,m) occurs in $(n-2)!/((k-1)!2^{k-1})$ such sets ω.□

A consequence of the theorems 3.8 and 3.9 is

Theorem 3.10. Let a sequence $\{A_j\}$ starting with $A_0 = A$ be constructed recursively by

$$A_j = (P_jQ_jV_j)^{-1} A_{j-1} P_jQ_jV_j,\ j = 0,1,\ldots$$

where in each step k disjunct index pairs $(\ell_1^j,m_1^j),\ldots,(\ell_k^j,m_k^j)$ are selected according to rule (3.20). P_j is a permutation matrix with $P_j(\ell_1^j,m_1^j,\ldots,\ell_k^j,m_k^j)^T = (1,2,\ldots,n)^T$, Q_j is a preconditioning orthogonal block-diagonal matrix as described in this section and $V_j = \mathrm{diag}(x_j, x_j^{-1},\ldots,x_jx_j^{-1})$ reduces the Euclidean norm of $(P_jQ_j)^{-1}A_{j-1}P_jQ_j$ as described in theorem 3.8. Then $\{A_j\}$converges to normality.

Proof. Since $\Delta_j = \|A_{j-1}\|_E^2 - \|A_j\|_E^2 \to 0\ (j\to\infty)$ and, as follows from theorem 3.8 and theorem 3.9.

$$\Delta_j \geqq \frac{1}{8}\|A\|_E^{-2}\sum_{\ell=1}^{k}\|v^j\|^2 \geqq \frac{1}{2(n-1)}\|A\|_E^{-2}\|C(A_{j-1})\|_E^2 ,$$

we conclude $C(A_j)\to 0(j\to\infty)$.□

Remark 3.11. Evidently the same choice for P_jQ_j together with the optimal norm-reducing $V_j = \text{diag}(S_1^j S_2^j,\ldots,S_k^j)$ where $S_\ell^j = S^j$, $\ell = 1,\ldots,k$ as described in theorem 3.5 provides a sequence $\{A_j\}$ that so much the more converges to normality.

Finally we mention that each S_ℓ^j is row congruent with the shear $S_\ell^j U_\ell^j$ that diagonalizes the symmetric part of the current matrix. Veselić (1976) proved that a sequence of norm reductions interrupted by Jacobi iterations for the diagonalization of $A_j + A_j^T$ effects that $\lim_{j\to\infty} A_j = D + K$, with $D = \text{diag}(\text{Re}(\lambda_j))$, $K = -K^T$ and $DK = KD$. Then $D_{ii} \neq D_{jj}$ implies $K_{ij} = 0$.□

Remark 3.12. For concreteness we indicate the parallelization of a cyclic version of the norm-reducing process with the caterpillar permutation P. Then the time-consuming search in (3.20) is avoided.

(i) Annihilate the elements $c_{2\ell-1,2\ell}$, $\ell = 1,\ldots,k$ of C(A) with $A' := Q^T(AQ)$. The updating of the column pairs $2\ell-1, 2\ell$ can be performed simultaneously: $A \to A\,\text{diag}(Q_1,\ldots,Q_k)$; once this is done the updating of the row pairs $2\ell-1, 2\ell$ can be carried out concurrently: $AQ \to \text{diag}(Q_1^T,\ldots,Q_k^T)(AQ)$ with k processors $G_1,\ldots,G_k$,

(ii) Compute with processor G_ℓ:

$$e := \sum_{m=1}^{k} (a'_{2\ell-1,2\ell})^2, \quad f_\ell := \sum_{m=1}^{k} (a'_{2\ell,2m-1})^2$$

Let be $E := \sum_{\ell=1}^{k} \ell_\ell$, $F = \sum_{\ell=1}^{k} f$ and

$$x = \begin{cases} 1 & E = F = 0 \\ (1-\frac{1}{4}\|A\|_E^{-2}\,E)^{1/4} & , \quad E \neq 0,\ F = 0 \\ (1-\frac{1}{4}\|A\|_E^{-2}\,F)^{-1/4} & , \quad E = 0,\ F \neq 0 \\ (E/F)^{1/4} & , \quad E,F \neq 0 \end{cases}$$

The column and row updates $A' \to \tilde{A} := V^{-1}A'V$ with $V = \text{diag}(x,x^{-1},\ldots,x,x^{-1})$ can be performed concurrently as in (i).

(iii) Execute the caterpillar permutation $A := P^TAP$.

After n-1 of these steps the original order has been restored. The analysis of this parallel process already leads to many problems of design and it makes the clear the importance and difficulties of the data flow and communication in the implementation of parallel algorithms.

REFERENCES

Brent, R.P., and Luk, F.L., 1985, The solution of singular-value and symmetric eigenvalue problems on multiprocessor arrays, SIAM J. Sci. Stat. Comput., 6:69-84.

Dollinger, E., 1981, Ein linear konvergentes zyklisches Jacobi ähnliches Verfahren für beliebige reele Matrizen, Num. Math., 38:245-253.

Eberlein, P.J., 1962, A Jacobi-like method for the automatic computation of eigenvalues and eigenvectors of an arbitrary matrix, J. Soc. Industr. Appl. Math., 10:74-88.

Elsner, L., and Paardekooper, M.H.C., On measures of non-normality of matrices, Lin Alg. Appl., 92:107-124.

Forsyth, G.E., and Henrici, P., 1960, The cyclic Jacobi method for computing the principal values of a complex matrix, Trans. Am. M.S., 94:1-23.

Goldstine, H.H., and Horowitz, L.P., 1959, A procedure for the diagonalization of normal matrices, J. Assoc. Comp. Mach., 6:176-195.

Hari, V., 1982, On the quadratic convergence of the Paardekooper method I, Glasnik Matem., 17:183-195.

Hari, V., 1982, On the global convergence of the Eberlein method for real matrices, Num. Math., 39:361-369.

Hestenes, M.R., 1958, Inversion of matrices by bio-orthogonalization and related results, J. Soc. Indust. Appl. Math., 6:51-90.

Jacobi, C.G.J., Uber ein leichtes Verfahren die in der Theorie der Seculär störungen vorkommenden Gleichungen numerisch aufzulösen, Crelle's J., 30:51-94.

Kempen, H.P.M. van, 1966, On the convergence of the classical Jacobi method for real symmetric matrices with non-distinct eigenvalues, Num. Math., 9:11-18.

Kempen, H.P.M. van, 1966, On the quadratic convergence of the serial cyclic Jacobi method, Num. Math., 9:19-22.

Luk, F.T., 1980, Computing the singular-value decomposition on the ILLIAC IV, ACMTOMS, 6:524:539.

Modi, J.J.,and Pryce, J.D., 1985, Efficient implementation of Jacobi's diagonalization method on the DAP, Num. Math., 46:443-454.

Paardekooper, M.H.C., 1969, An eigenvalue algorithm based on norm-reducing transformation, Technical University Eindhoven, Thesis.

Paardekooper, M.H.C., 1971, An eigenvalue algorithm for skew symmetric matrices, Num. Math., 17:189-202.

Paardekooper, M.H.C., 1986, Sameh's parallel eigenvalue algorithm revisited, Research Memorandum, Tilburg University.

Ruhe, A., 1968, On the quadratic convergence of a generalization of the Jacobi method for arbitrary matrices, BIT, 8:210-231.

Rutishauser, H., 1966, The Jacobi method for real symmetric matrices, Num. Math., 9:1-10.

Sameh, A.H., 1971, On Jacobi and Jacobi-like algorithms for a parallel computer, Math. Comput.,25:579-590.

Schönhage, A., 1964, On the quadratic convergence of the Jacobi process, Num. Math., 6:410-412.

Stewart, G.W., 1985, A Jacobi-like algorithm for computing the Schur decomposition of a non-hermitian matrix, SIAM J. Sci. Stat. Comput., 6:853-864.

Veselić, K., 1976, A convergent Jacobi method for solving the eigenproblem of arbitrary real matrices, Num. Math. 25: 179-184.

Wilkinson, J.H., 1962, Note on the quadratic convergence of the cyclic Jacobi process, Num. Math. 4:296-300.

POSTSCRIPTUM

Since the 1985 Workshop various extensions of the basic Jacobi algorithm were presented. They include an alternative parallel Schur decomposition algorithm (Eberlein, 1987), a construction of the "closest normal matrix" (Ruhe, 1987) and a Jacobi-like parallel annihilation method for nonhermitian diagonal dominant matrices (Paardekooper, 1987). The two-sided block Jacobi method for the singular value decomposition (Bisschof, 1987) has been powerfully implemented on a hypercube.

REFERENCES

Bisschof, C., 1987, The two-sided block Jacobi method on a hypercube, _in_: "Hypercube Multiprocessors", M. Heath, ed., SIAM, 612:618.

Eberlein, P.J., 1987, "On the Schur decomposition of a matrix for parallel computation, _IEEE Trans. Computers_ C-36: 167-174.

Paardekooper, H.H.C., 1987, A quadratically convergent parallel Jacobi process for diagonal dominant matrices with distinct eigenvalues, Research Memorandum, Tilburg University.

Ruhe, A., 1987, Closest normal matrix finally found, _BIT_ 27: 585:598.

HIGH PERFORMANCE COMPUTING IN ECONOMICS

H.M. Amman*

Department of Macroeconomics, Faculty of Economics
University of Amsterdam, Jodenbreestraat 23, NL-1011 Amsterdam, The Netherlands

ABSTRACT

In this paper it is briefly pointed out to what extent recent developments in supercomputers can facilitate scientific computation in the various fields of economics. With the help of an example of adaptive control simulation it is made clear that with the current computation speed available, the possibilities for economic science are increased considerably and give rise to new fields of applications.

1. INTRODUCTION

In the last few years there has been a tremendous progress in computer architecture. An example of this development is a new generation supercomputer capable of performing high speed computations on large data structure. It is therefore no wonder that supercomputers are of special interest to scientists confronted with large scale scientific computational problems. In fact, supercomputers are mainly used for scientific purposes and only to a small degree in other fields like industry and for image processing. In physics, chemistry and engineering more and more attention is paid to the increased possibilities offered by supercomputers. Tue to their high speed of computation, supercomputers are considered as a major tool to push the boundaries of science.

Although in economics quite similar computational problems exist as in other sciences, up to now, little or no attention has been paid to this potential field of research. One of the

---

* I would like to thank Professor David Kendrick, University of Texas at Austin, for his encouragement and making his DUAL-code available for implementation on a number of supercomputing machines. I am indebted to Lidwin van Velden, Dirk Kabel and Rob van Houweling for their assistance and comments on an earlier version of this paper. Supercomputer access was financed by the Dutch National Science Foundation (NWO), IBM Nederland and the Dutch Space Laboratories (NLR).

few exceptions (Amman, 1985; Ando et al., 1987; Foster et al.; Henoff and Norman, 1986 and Petersen, 1987). Subsequently it is interesting to look at the new opportunities that supercomputing can offer for economic science.

This paper is organized as follows. In Section 2 it is pointed out in what fields of economics supercomputing can be of importance. Section 3 presents an example and some benchmarks of a computational problem in economics which is still considered as a 'hard core' computational problem. Finally, Section 4 summarizes the main conclusion of this paper.

2. FIELDS OF APPLICATIONS IN ECONOMICS

In general, we could say that supercomputing only pays off for computationally intensive and I/O intensive problems. Commonly, scientific computational problems are characterized by both these elements. The large number of floating point operations are very often the result of nonlinearities in the models employed. These nonlinearities necessitate the use of an iterative procedure which means that the number of iterations is large, and subsequently, the number of floating point operations is large. More specific for economics is the problem of data motion. Economic models are large compared to models in Physics and Engineering. In order to simulate with these models we have to deal with large data sets. This means that the I/O time connected to this type of simulation is major. Supplementary, economists are not interested in "a" solution of a model but in a specific solution: the "best" or optimal solution of a model. So in fact, most models in economics are actually optimization models instead of mere models describing the "law of economics". It is evident that this optimality condition adds an additional degree of complexity to the "numerics" of the problem.

It is clear that the use of a supercomputer is justified for only a special class of problems. Computational problems in economics that are possible candidates for supercomputing are

- Macroeconomic policy evaluation
- Nonlinear estimation techniques for large systems
- Financial modelling
- Optimal control simulations

The models in economics used for policy evaluation purposes are generally large dynamic, linear or nonlinear, models that have to be solved for a certain time span. Model sizes up to 1000 equations are no exception. There are even examples of large models, like the LINK-model, with more than 10^5 equations. Solving these models is, due to their size, computationally intensive. For this type of problems a supercomputer makes it possible to generate model solutions in a reasonable processing time and when the model is not too large it is possible to produce results in a real-time fashion. An example of this type of usage on an IBM 3090/VF is presented in Petersen (1987).

A second application field is nonlinear estimation techniques. Nonlinear estimation techniques, like Full Information Maximum Likelihood (FIML), for simultaneous equations models are computationally intensive and are carried out on large data sets. For this reason the memory requirements are large and will result in a long computation time.

A very recent application is the modelling in connection with financial markets like the option market, stock market and the bond market. One of the main characteristics of financial markets is that changes in these markets can take place very rapidly. Black Monday (October 19, 1987) has once again shown that in a very short ime, prices may change significantly. In order to respond adequately on changes in the market it is necessary for large market participants, like brokers and banks, to evaluate information as fast as possible. This is necessary in order to make, for instance, the appropriate portfolio changes. Any delay in information access would mean a financial loss.

According to the methodology developed by Black and Scholes (1973), it is possible to approximate the implication of market changes on portfolio positions. Their method requires, however, that a set of partial differential equations has to be solved by means of numerical integration. Depending on the number of assets taken into account, obtaining a solution can be highly computationally intensive. For this application supercomputers can make it possible to have online information processing, making instantaneous response possible. An example for the bond market can be found in Foster et al.

The fourth and last application field that currently can be distinguished, is the use of supercomputers for optimal control simulations. In this application field model simulations are carried out conditional on some form of "dynamic optimality". This means that for a certain time period a performance index is optimized. Normally, these simulations are performed in a deterministic setting applying a deterministic economic model. The problem becomes more complicated, however, if the optimal control simulations are carried out in a stochastic fashion. In the next section we will discuss such an example.

3. AN EXAMPLE

In earlier papers (Amman, 1985; Amman, 1986 and Amman, 1988) we investigated the possibilities of supercomputers for control simulations on deterministic systems. A logical step forward is to extend the analysis for stochastic systems. There are a number of mathematical procedures for modelling stochastic control systems, which are limited in their applicability due to computational difficulties. For instance, adaptive control experiments cause severe computational difficulties. Following Kendrick (1981), an adaptive control problem can be formulated as

$$\min: J_N = E\{C_N\} \tag{1}$$

where

$$C_N = L_N(x_N) + \sum_{k=0}^{N-1} L_k(x_k, u_k) \tag{2}$$

Subjected to the conditions

$$x_{k+1} = f_k(x_k, u_k) + \mu_k \tag{3}$$

and the requirement equations

$$y_k = h_k(x_k) + \Omega_k \tag{4}$$

Additionally we assume that

$$E\{x_o\} = \hat{x}_{o|o} \qquad cov(x_o) = \Sigma_{o|o}$$

$$E\{\mu_k\} = 0 \qquad cov(\mu_k) = Q_k$$

$$E\{\Omega_k\} = 0 \qquad cov(\Omega_k) = R_k$$

where

x = n element state vector
u = m element control vector
y = r element measurement vector
L(.) = scalar performance index
Q,R = covariance matrix
μ,Ω = gaussian random terms

using k as a time index. Conform to Tse, Bar-Shalom and Meier (1973), Kendrick solves the Closed Loop form of the above adaptive control model by means of an approximation along a nominal path. In the first stage the certainty equivalence (deterministic) part is solved using linear quadratic control techniques and in a second stage the so-called optimal costs to go are determined by means of a forward objection. In performing the forward projection a distinction is made between the certain equivalence part, cautious and a probing part of the solution. Additionally, at every time state the coefficients of the model are reestimated with the help of kalman filtering.

Due to this forward projection and reestimation the problem is computationally intensive even if the model is relatively small. Until now, this type of technique could not be used for real life models. With the new generation supercomputer machines however, new possibilities come into account. To get insight into the possibilities of supercomputing machines for the above adaptive control problem we have implemented an existing code, Kendrick's DUAL-code, on a number of supercomputing machines. The DUAL-code was implemented on a NEC SX/2, a Cyber 205, a Cyber 995, and an IBM 3090-180 with and without a vector processor. As these machines are basically (SIMD) vector machines the DUAL-code was vectorized according to the familiar rules described in, for instance, the Control Data Corp. Fortran 200 Reference Manual, Dongarra and Eisenstat (1984), Dongarra et al. (1984), Dongarra and Hewitt (1986), Dongarra (1987), Hoffman (1987) and Johnston (1985).

In order to carry out some realistic benchmarks, we employed for the system equations (3) a medium sized 'real life' econometric model of 75 state variables, 7 control variables, 5 unknown coefficients and 10 measurement equations§. The model was used to make a three-month (one period) prediction. The

§ For a description of the model see Amman and Jager (1987).

Table 1. Benchmarks[1] of the dual code for one period run

Computer system	CPU hours	Vector Pipes	Max. mflop[2] performance
Cyber 995E	53.168	2	134
Cyber 205	31.207	2	200
IBM 3090-180	28.854	-	14
IBM 3090-180/VF	22.533	1	119
NEC SX/2	2.642	4	1300

[1] All benchmarks in 64-bit precision.
[2] Peak rate measured in mflops.

IBM and NEC machines use a standard 32-bit precision while the Cyber machines use a 64-bit precision. As for the adaptive control experiments the level precision turned out to be important, control experiments were performed in 64-precision. Precision turned out to be especially important for the convergence termination condition. At a 32-bit precision, level loss of significance occurred resulting in premature termination. Consequently, the computations on the IBM and NEC were carried out in double precision, 64-bit precision.

Table 1 clearly indicates that adaptive control experiments require a lot of CPU when applied to a model of realistic size. When utilizing a machine like the NEC SX/2 the solution is obtained in an acceptable turn-over time. What is furthermore interesting, is that the relative performance of the scalar machine, the IBM 3090 without vector facility, is not dramatically lower compared to the vector machines. This demonstrates that for the dimensions of the model we used, vector mode is only moderately more efficient than scalar mode. This typically occurs when the average vector operated on is of insufficient length leading to a deterioration of vector processing. If the so-called $n_{\frac{1}{2}}$ condition is not fulfilled (cf. Hockney and Jesshope, 1987)) it is very likely that the time needed to start up a vector operation is not compensated by an adequate vector length. From this one can conclude that for the model size vector processing features are not enough to deal with the computational complexity. This strongly advocates for a further exploitation of parallel processing features for our type of control algorithm.

4. CONCLUSIONS

From the previous sections it has become transparent that with the current state of technology new methods can be applied and larger computational problems can be tackled. Computational problems in economics that were considered as nonsolvable until recently can now be dealt with. The supercomputers presently available and those that will become available in the near future, are offering a powerful tool for economic and econometric analysis. For this reason it is to be recommended to expand effort and time in the field of supercomputing in economics.

5. REFERENCES

Amman, H.M., 1985, Applying the Cyber 205 for optimal control experiments in economics, Supercomputer 8/9:71-74.

Amman, H.M., 1986, Are supercomputers useful for optimal control experiments?, Journal of Economic Dynamics and Control 10:127-130.

Amman, H.M., and Jager, H., 1987, Optimal economic policy under a crawling-peg exchange-rate system, in: "Developments in Control Theory for Economic Analysis", C. Carroro and D. Sartore, eds., Martinus Nijhoff, Deventer, p. 105-125.

Amman, H.M., Nonlinear control on a vector machine, Parallel Computing, forthcoming.

Ando, A., Beaumont, P., and Ando, M., 1987, Efficiency of the Cyber 205 for stochastic simulation of a simultaneous nonlinear dynamic econometric model, International Journal of Supercomputers and their Applications.

Black, F., and Scholes, M., 1973, The pricing of options, corporate liabilities, Journal of Political Economy 81:637-654.

Control Data Corporation, 1982, Fortran 200 Reference Manual, Publication number 60480200.

Dongarra, J.J., and Eisenstat, S.C., 1984, Squeezing the most out of an algorithm in Cray Fortran, ACM Transactions on Mathematical Software 10:219-230.

Dongarra, J.J., Gustavson, F.G., and Karp, A., 1984, Implementing linear algebra algorithms for dense matrices on a vector pipeline review, SIAM Review 26:91-112.

Dongarra, J.J., and Hewitt, T., 1986, Implementing dense linear algebra algorithms using multitasking on the Cray X-MP-4 (or approaching the Gigaflop), Siam Journal on Scientific and Statistical Computation 7:347-350.

Dongarra, J.J., 1987, Performance of various computers using standard linear equations software in a Fortran environment, Argonne National Laboratory, Technical Memorandum No. 23.

Foster, G., Foote, W., and Kreamer, J., Force: Financial Options - Rapid Calculations and Execution, Mimeo Syracuse University, New York.

Henoff, P., and Norman, A.L., 1986, Solving nonlinear econometric models using vector processors, in: "Proceedings of the IFAC Workshop on Modeling Decision and Games with Applications to Social Phenomena".

Hockney, R.W., and Jesshope, C.R., 1987, "Parallel Computers", Adam Hilger, Bristol.

Hoffmann, W., 1987, Solving linear systems on a vector computer, Journal of Computational and Applied Mathematics 18:353-367.

Johnston, C.H.J., 1985, Matrix arithmetic on the Cyber 205, Supercomputer 8/9:28-48.

Kendrick, D.A., 1981, "Stochastic Control for Economic Models", McGraw Hill, New York.

Petersen, C.E., 1987, Computer simulation of large scale models, project link, in: "Scientific Computing on IBM Vector Multiprocessors", R. Benzi and P. Sguazzero, eds., Volume II, ECSEC.

Schönauer, W., 1987, "Scientific Computing on Vector Computers", North-Holland, Amsterdam.

Tse, E., Bar-Shalom, Y., and Meier, L., 1973, Wide sense adaptive control for nonlinear stochastic systems, IEEE Transactions on Automatic Control 18:98-108.

van der Vorst, H.A., and van Kats, J.M., 1984, The performance of some linear algebra algorithms in Fortran and Cray-I and Cyber 205 supercomputers, University of Utrecht, ACCU Technical Report TR-17.

BENCHMARK ON SUPERCOMPUTERS FOR AN INDUSTRIAL ENVIRONMENT

E. Vergison

Solvay Scientific Computer Centre

rue de Ransbeek 310, B-1120 Brussels, Belgium

ABSTRACT

The aim of this study is to provide an industrial scientific and technical computer center with a reliable and exhaustive tool for making comparisons of performances between different computers including those of the new generation (parallel and vector machines).

1. INTRODUCTION

The methodology adopted for providing an industrial scientific and technical computer center with a reliable and exhaustive tool for making comparisons of performances between scalar and vector or parallel computers, is based on three main approaches:

- general purpose considerations covering scalar and vector compilation, computer architecture, programming language, debugging, development tools and training and education.

- practical test cases, themselves subdivided into three subgroups:

    - basic operations and manipulations such as floating point operations, elementary mathematical functions, do-loops and branching.
    - linear algebra, this chapter being of major use for the practical applications and allowing precise estimates of the computational work involved. This chapter covers vector and matrix manipulations and direct algebraic linear system solvers.
    - industrial applications covering fluid mechanics, chemical engineering, heat transfer and management.

- complementary studies in order to measure the manpower investment needed to make an intelligent use of the vector facility, either in vectorizing existing code, or by developing automatic tools capable of overcoming vector inhibitors.

The test cases under consideration were programmed in different ways in order to show what working principles the vectorizing processors are based on, and to highlight the compiler capabilities as well as the programming effort that is to be expected.

The language used is full FORTRAN 77/ANSI X3.9 - 1978 (only a few tests were carried out in FORTRAN 66).

The computer environments on which the tests were carried out, were composed of two subgroups of machines: mini-computers and super-computers, keeping in mind both departmental and centralized solutions.

A summary of the machines' overall performances is presented in the Appendix.

2. THE "CLASSICAL" BENCHMARK

Since we aimed to check both scalar and vector capabilities of the machines in order to obtain an estimate of the absolute performances of computer codes, the benchmark is CPU oriented and all the tests were carried out in a single user environment.

Although we focused especially on computing performances, the benchmark was also designed to study compiled Fortran code efficiency and the ability of compilers to vectorize, that is, to render programs automatically suitable for super performance.

2.1. General Purpose Information

To begin with, we need a comprehensive view of what parallelism or vectorization means for each manufacturer. So, before starting any machine test, we have to have clean and comprehensive documentation on the following subjects:

Architecture:

- to which category does the computer belong: pipelined, array processor, MIMD [1]?
- how does vector processing work, i.e. how does it process array elements?
- how does the system handle multitasking, if any?

Compiler:

- what about compiler capabilities, what does it vectorize and how?
- are there levels of vectorization; for instance does the compiler handle individual statements inside do-loops? If any, which ones?
- what about compiler versatility? Can we compile both in scalar and vectorized mode the programs written in scalar Fortran 77?
- what is its compatibility with Fortran 66? Does the program have to conform to Fortran 77 to execute correctly when vectorized?

- what are the compiler vectorizing optimization levels, if any?

Language:

- to what extent are there vector language extensions?
- if any, do they conform to the Fortran 8x draft?
- which statements are vectorizable?
- which data types can be handled by the vectorizing compiler?

Performance Analyzers (Profilers):

- to what extent does the system provide the user with runtime information like the time spent in specified parts of the code or in subroutines?

Reporting:

- to what extent does the system produce a cross reference map or an output that shows the way vectorization was performed and to what extent does it give directives to improve the coding?

Debugging:

- can programs be debugged at any optimization or vectorization level?

Mathematical support:

- does the manufacturer provide the user with any optimized mathematical software?
- to what extent does it conform to the Basic Linear Algebra Subroutines concepts?

Third party software:

- does the machine allow for widely used scientific and engineering software?

2.2. Practical Testcases

All these tests are programmed in standard Fortran 77 and the typical vector sizes used are 50, 64, 100, 300, 500 and 10,000.

Basic arithmetic and programming operations: these include floating point operations and elementary mathematical functions evaluations on vectors, do-loops and branching.

Floating point operations: sixteen cases, taken from a benchmark by Karaki [2], are considered; they reflect common manipulations of scientific programming.

Elementary mathematical functions: frequently used mathematical functions have been tested. The only point to be stressed is the choice of different domains of definition for the sine, cosine and tangent function in order to take into account the fact that computing algorithms may depend on it.

Do-loops: several scenarios involving this typically well suited structure for vectorization are considered [3]: flow dependence, anti dependence, vectorization by loop distribution and reordering, partial vectorization by loop distribution, as well as reducing scalar dependence by scalar manipulations.

Branching: this Achilles' heel of vectorization is, a priori, a hard to vectorize structure. Nonetheless, we tried to detect possible parallelism by using Boolean structure and a masking technique [3].

2.3. Linear Algebra

Four groups of problems have been tackled:

- vector manipulations including the dot product, the DAXPY form and vector components summation
- matrix by vector products based either on the dot product or on the DAXPY form
- matrix by matrix multiplication based on the same techniques and
- direct linear algebra solvers.

Vector manipulations:

The dot product $\Sigma X_i Y_i$, $i = 1,\ldots,n$ and the DAXPY form $Z_i = Y_i + aX_i$, $i = 1,\ldots,n$ are the basic structures of many algorithms in linear algebra. Their treatment depends strongly on the hardware architecture, so we found it interesting to check their respective performances as BLAS level 1 operations.

Matrix by vector product:

The performances of both the dot product based multiplication:

```
    d0 10 i = 1,n
    d0 10 j = 1,n
 10 c(i) = c(i) + a(i,j) * b(j)
```

or the DAXPY form one:

```
    d0 10 i = 1,n
    d0 10 j = 1,n
 10 c(j) = c(j) + a(j,i) * b(i)
```

have been coded.

Matrix by matrix product:

Here again, dot matrix and DAXPY forms were implemented as well as the "outer product" algorithm

```
    d0 10 k = 1,n
    d0 10 j = 1,n
    d0 10 i = 1,n
 10 c(i,j) = c(i,j) + a(i,k) * b(k,j)
```

A point of supplementary interest concerning these matrix by matrix routines is the way the vectorizing compiler handles nested do-loops.

Algebraic linear systems:

The purpose of this chapter is to solve linear algebraic systems using direct methods: the Gaussian and the Crout variant for non-symmetric matrices and the Cholesky for symmetric ones.

Both the Crout and Gauss methods require the same computational effort, $(2/3)n^3$ operations, n being the matrix dimension) while Cholesky's needs only half of this. For example, a 100 x 100 non-symmetric system would require about 680,000 floating point operations (flops).

The matrix under study is the Frank one that takes the form

$$(a(i,j))_{nxn} = (n+1 - \max(i,j))_{nxn}$$

It has an interesting feature being the inverse of a matrix which can be concretized as a discretized form of a two point boundary value problem.

Two kinds of problems were implemented: a non-sophisticated one (diagonally dominant pivoting) for both the Gauss and the Crout methods, the overheads being minimized and a more robust form of both the Gauss and the Cholesky methods coded in a BLAS form by Dongarra and Eisenstat [4].

2.4. Application Programs

The applications described in this chapter are divided into four groups. For the first one, we can reasonably expect improvement simply by compiling; programs in the second need changes at the algorithmic level. The third group is made up of programs which potentially contain parallelism, whilst the last one contains those where parallelism can hardly be expected a priori. To be closer to reality, some programs are written in single precision.

Applications with compiler detectable parallelisms:

Two applications have been selected, one in fluid mechanics and one in heat transfer.

Fluid mechanics: a steady two-dimensional turbulent fluid flow is calculated, using an iterative "time marching" technique based on the finite difference Marker and Cell (MAC) method. This program, implemented in single precision, was dealt with in further complementary studies [6].

Potential equation: a potential equation solver, based on a multigrid algorithm and written in double precision, was added to the benchmark because the multigrid technique offers three interesting features:

- it is one of the most powerful PDE solvers
- it is naturally vectorizable
- it is easy to make operation count estimations.

Application programs needing algorithmic changes for vectorization:

A one-dimensional nonlinear heat equation, using an unconditionally stable algorithm requiring a tridiagonal linear system solver, has been coded in two ways (both in double precision). In the first version, the tridiagonal system is solved by the classical Gaussian elimination and backward substitution. A total of 9n scalar arithmetic operations is required, but we have to keep in mind that there is one division and some overhead due to integer manipulations as well as integer.

A variant, based on a cyclic reduction algorithm, better suited for parallel computing, has been implemented. It requires $17\ (n + n_{\frac{1}{2}} * \log_2 n)$ operations where $n_{\frac{1}{2}}$ is the machine half performance length [1], typically 10 for a CRAY 1, 2 for a PPS/164 and 0 for a scalar machine.

Potentially vectorizable applications:

Many applications in chemical engineering use numerical algorithms which can be viewed as fixed point iteration techniques (Picard's method); these algorithms are "vector minded" and should give good results on parallel machines.

First a nonlinear regression analysis program designed to check the statistical consistency of productions and consumptions by finding the maximum likelihood estimates of the model parameters was coded in Fortran 77, in double precision. The constrained nature of the problem introduces non-linearities that are handled using Picard's method.

Secondly, a set of kinetic equations, solved using the classical Runge-Kutta method was implemented. Due to problem complexity, the formulation of the second members of the differential equations makes an intensive use of function cells which can actually prevent any benefit of vectorization. This test case bridges the gap with the next paragraph.

Applications where parallelism is not to be expected:

As far as we know from our experience, the programs which fall into this category are the operations research ones, those using complex indirect addressing in their numerical algorithms and those making an intensive use of subroutines and function calls. Two examples were used.

Operations research: A branch and bound algorithm with binary variables (0-1) and pure integer data, for solving a large linear system with linear constraints was implemented in single precision.

Linear mean squares: The minimal norm problem is to be solved: minimize

$$\| Ax - b \|_2,$$

where A belongs to $R^{m \times n}$, $m \geqslant n$ and is sparse with random structure. Sparsity here means that the matrix density d(a), defined as

$$d(a) = \frac{\text{number of nonzeroes of A}}{m \times n}$$

should be less than or equal to 0.05. QR factorization exploring sparsity at symbolic level, was the algorithm adopted for solving this problem. It was implemented in Fortran 77, in double precision, for the floating point aspects, otherwise in integers.

3. THE COMPUTER ENVIRONMENTS

We first recall that the tests were carried out in a single user environment and that the set of tested machines has

The computer environments of the benchmarks.

CRAYETTES	
FPS M64/60	Attached processor & mono-user Wide instruction word machine SJE operating system
CONVEX C1/XP	Stand alone & multi-user Vector machine UNIX derived operating system
CONVEX C210	Stand alone & multi-user Vector machine UNIX derived operating system
ALLIANT FX/4	Stand alone & multi-user Parallel machine with 4 computer elements UNIX derived operating system
ETA10 MOD P	Stand alone & multi user Vector machine EOS operating system
SUPERCOMPUTERS	
CYBER 990	Stand alone & multi user Vector machine NOS/VE operating system
CYBER 205	Stand alone & multi user Vector machine NOS/VE operating system
IBM 3090/150	Stand alone & multi user Scalar machine with vector facility MVS or VM/CMS operating system
IBM 3090/180	Stand alone & multi user Scalar machine with vector facility MVS or VM/CMS operating system
CRAY/XMP 145E(*)	Back-end processor Vector machine COS operating system

(*) by courtesy of Prof. R. Devillers from Brussels Free University.

been divided into two subgroups: the mini-supercomputers or "crayettes" and the supercomputers as shown in the table (the scalar DEC 8700 was chosen as a reference).

The tables in the Appendix summarize the results obtained on all the machines except the IBM 39090/180 and the Cyber 205, the reason being that:

- the overall performances of the IBM were estimated at 1.2 times those obtained from the IBM 3090/150. We also note that we were interested in the IBM 3090/180 machine because it is the basic module of the IBM 3090 field upgradable family (models 200 to 600)

- the results we obtained from the CYBER 205 are estimated at 2.5 times those we got from the CYBER 990.

We will now comment on these results.

Results from the Basic Operations

The results, presented in Table 1 of the Appendix, are average values with typical vector lengths of n = 50 and n = 500, which were considered as the most representative of our work. Detailed results are shown in [4].

Results from Linear Algebra

In Table 2 of the Appendix, the dot product and DAXPY form results are presented for typical vector sizes of 64 and 512.

Tables 3 and 4 show how the dot product and the DAXPY form reflect when implemented in both the matrix by vector and the matrix by matrix multiplication. Of great interest is the do-loop unrolling technique (Table 4) which reduces memory references by using the vector registers more efficiently [5].

The benchmark results of linear algebraic solvers in Table 5 of the Appendix, were completed by some machine specific equivalents. The use of these specially tuned routines generally improves performances substantially.

Results from the Applications

One should be very careful when interpreting the results summarized in Table 6 of the Appendix. We must remember that:

- the application part of the benchmark was set up without any special tuning, even if some applications were, a priori, vectorizable and if some additional work was carried out afterwards either by the manufacturer or by ourselves

- though most applications were programmed using double precision, three important exceptions were included in the benchmark: a fluid dynamic code written in single precision, an operations research one dealing with integer programming and a linear mean square solver mixing double precision and integer programming

- a tuning effort was carried out on two programs: the non-stationary heat solver and the fluid dynamic code [6]. One general observation is that, without any changes in the codes, only the supercomputers actually improve the performances of the whole set of programs. However, all of them do it significantly after some programming or algorithmic work.

Particularly impressive is the speedup already obtained by the replacement of old Fortran 66 structures by Fortran 77 structures, and by inserting frequently called routines inline [6].

Overall Performances

Although overall performances naturally raise objections and criticisms, we found it appealing:

- to summarize in one table the average ratios we got from the computations, the DEC 8700 being taken as a unit

- to give, to some extent, a realistic although somewhat conservative idea of the machine's efficiency.

Table 7 in the appendix summarizes the characteristics of the machines tested so far. By "presently expected ratios" we mean that weights were introduced to take into account the fact that industrial libraries contain programs that are, a priori, not vectorizable or that are not written in a modern structured language.

4. COMPLEMENTARY STUDIES

Two studies, complementary to the benchmark described earlier, were conducted in order to quantify the manpower needed to vectorize a fluid dynamics code and to developed an inline routine inserting pre-compiler in order to get round subroutine calls which, used in do-loops, seriously inhibit vectorization.

Adaptation of a Fluid Dynamic Code

The objective of this work, conducted in collaboration with Brussels Free University [6], was to vectorize the fluid dynamic code we mentioned in "Applications with Compiler Detectable Parallelism".

The procedure that was followed, can be divided in four steps:

- performance analysis
- scalar re-programming
- algorithmic changes
- vector programming

Performance analysis:

A profiler (performance analyzer) was used to detect those parts of the code that are the most time consuming. This information coupled with relevant directives provided by efficient compilers showed where to put one's effort. In our example, one routine consumed 60% of the total CPU time and another one 20%.

Scalar re-programming:

We observed, when re-coding those parts of the code detected by the profiler, that inline insertion of routines, for example, already improved scalar performances. Also replacing branching structures by masking led to the same conclusion. These changes are to be considered as a pre-conditioning of the code.

Algorithmic changes:

These may either affect the whole algorithmic strategy of the problem or only part of it. In the fluid dynamic code under consideration the time marching procedure chosen to reach the steady state reflects an explicit iteration technique which does not need special adaptation.

It was not the case for the innermost pressure solver where the line by line mesh sweeping had to be replaced by red-black ordering.

Vectorization:

The first transformation that was made, was to re-write matrices as long vectors applying gathering techniques in order to avoid space wasting and to allow fast access to equally spaced vector components.

The second one consisted in inserting parentheses to force execution of the arithmetic operations in the most efficient order (linked triads for example).

Summary of results of code vectorization:

The following speedup ratios are calculated, the DEC 8700 being taken as a unit:

	Before tuning	After tuning
CONVEX C1 XP	0.7	6.1
CONVEX C210	-	16.8
FPS 64/60	3.1	7.9
ETA 10 MOD P	-	35.6
ALLIANT FX/4	0.7	6.5
IBM 3090/150	4.8	9.4
CYBER 990	3.2	5.5
CRAY XMP/14SE	17.3	19.0

In Line Inserting Pre-Processor

The procedure followed, in a study made in collaboration with Brussels Free University, is made up of two steps preceding the compile link-go: code standardization and source code pre-processing.

Code standardization:

In order not to make the pre-processing too complex, the source code is standardized using TOOLPACK, a set of programming development tools developed by the Numerical Algorithms Group [7].

Source code pre-processing:

This tool inserts routines in line,that means, replace subroutine calls by their full coding. The pre-processor must be told which routines to handle. This decision follows from the profiler's analysis. The compile link-go acts on the processed source code, while maintenance is done only on the original one.

5. CONCLUSIONS

In spite of the inherent limitations imposed on benchmarking work, amongst which:

- the necessity to make choices;
- the limited time available for benchmarking;
- the fact that the units used to quantify performances (seconds and MFLOPS) can be criticized, to some extent;
- the continuous improvements in hardware and software;
- the fact that machine workload and I/O's have not been taken into account,

we can conclude that we now dispose of a tool capable of:

- giving a very good idea of the vector and vectorizing capabilities and maturity of machines, both from a hardware and a software point of view, so reducing industrial choices to a few clear strategies;
- measuring precisely the effort needed to adapt old programs or to write new ones;
- checking how manufacturers cope with both third party software and the present standardization tendencies in scientific languages (Fortran 8X) and operating systems (UNIX);
- adapting easily to new approaches in computing science.

REFERENCES

1. R.W. Hockney and C.R. Jesshope, "Parallel Computers", Adam Hilger (1986).
2. Y. Karaki, Performance of mathematical libraries on the HITAC S-810 supercomputer, Supercomputer, May 1985, 37-46.
3. E. Vergison, Benchmark for scientific and technical computers in an industrial environment, Solvay Report, July 1987.
4. E. Vergison and T. Engels, Synthesis of the results from the Solvay scientific and technical benchmark on supercomputers, Solvay Report, July 1987.

5. J.J. Dongarra and S.C. Eisenstat, Squeezing the most out of an algorithm in CRAY FORTRAN, ACM Transactions in Mathematical Software 10:219-230.
6. F. Geleyn, Superordinateurs: test de performance et méthodologie de programmation, Student's work performed at the Brussels Free University, December 1987.
7. R. Iles, Toolpack/1 - version 2.1. - Quick reference guide, NAG Publication: NP1284, Oxford, October 1986.
8. I.N.R.I.A., Parallel Scientific Computation, Cours et Séminaires - Saint Cyprien (France), October 1985.
9. J.J. Dongarra, J. Du Croz, S. Hammerling and R. Hanson, A proposal for an extended set of Fortran basic linear algebra subprograms, Argonne National Laboratory, December 1984.
10. A. Lichnewsky, Parallelism and its use in vector computers, Course at ISPRA (Italy), June 1985.
11. J.J. Dongarra, Performance of various computers using standard linear equations software in a FORTRAN environment, Argonne National Laboratory, February 1986.
12. H. Foerster and K. Witsch, Multigrid software for the solution of elliptic problems on rectangular domains: MG00, Gesellschaft für Mathematik und Datenverarbeitung Mbh, Bonn, September 1982.

APPENDIX: SUMMARY OF THE BENCHMARK'S RESULTS

Table 1. Basic programming operations.

	DEC 8700	CONVEX C1 XP	CONVEX C210	FPS 64/60	ETA 10 MOD P	ALLIANT FX/4	IBM 3090/150	CYBER 990	CRAY XMP 14SE
FLOATING OPERATIONS (MFLOPS)	1.2	2.8	10.0	6.7	54.2	7.2	19.1	31.7	62.3
DO LOOPS (MFLOPS)	1.0	3.1	9.5	5.4	38.5	2.9	12.2	19.1	30.4
MATH. FUNCTIONS (MICRO SEC. PER OPERATION)	35.4	4.9	1.5	3.2	0.6	4.7	1.5	2.9	0.3
BRANCHING (MICRO SEC.)	580.	770.	286.	302.	638.	601.	127.	248.	393.

Table 2. Vector operations.

		DEC 8700	CONVEX C1 XP	CONVEX C210	FPS 64/60	ETA 10 MOD P	ALLIANT FX/4	IBM 3090/150	CYBER 990	CRAY XMP 14SE
DOT PRODUCT	64	1.3	9.1	12.7	11.5 / 14.6	25.1	4.3	15.6	25.6	22.9
	512	1.2	9.1	19.3	12.5 / 18.2	64.6	3.8	24.4	48.8	93.8
DAXPY FORM	64	1.2	9.0	12.2	8.9 / 10.1	106.1	2.8	8.4 / 13.8	32.0	75.3
	512	1.2	9.0	15.3	9.4 / 12.3	157.9	3.1	12.6 / 25.	41.0	82.0

Remark: when there are two figures in one cell, the second refers to tuned versions of the code.

Table 3. Matrix by vector product.

		DEC 8700	CONVEX C1 XP	CONVEX C210	FPS 64/60	ETA 10 MOD P	ALLIANT FX/4	IBM 3090/150	CYBER 990	CRAY XMP 14SE
DOT PRODUCT BASED	50	0.9	2.0	6.0	9.2	26.7	10.0	10.9	8.3	16.8
	500	0.5	2.9	8.4	11.9	116.4	1.4	3.6	5.3	52.6
DAXPY BASED	50	1.1	3.6	10.7	7.7	22.5	2.6	19.5	20.8	45.5
	500	1.0	5.5	15.2	9.4	104.9	2.7	22.7	38.8	79.6
UNROLLED FORM	50	1.1	5.2	-	13.4	21.2	5.0	7.5	8.8	60.7
	500	1.0	9.3	-	18.6	59.3	6.0	7.6	8.9	121.4
USING SPECIFIC LIBRARIES	50	-	12.6	38.1	29.7	-	-	48.4	-	-
	500	-	16.5	46.6	32.5	-	-	46.5	-	-

Table 4. Matrix by matrix product.

		DEC 8700	CONVEX C1 XP	CONVEX C210	FPS 64/60	ETA 10 MOD P	ALLIANT FX/4	IBM 3090/150	CYBER 990	CRAY XMP 14SE
BENCHMARK BEST FORMULATION	50	1.1	11.7	34.1	10.1	19.5	9.4	22.6	22.9	48.3
	300	1.0	15.5	42.6	11.8	71.5	7.7	21.0	36.8	75.4
USING SPECIFIC LIBRARIES	50	-	-	40.4	32.1	-	-	52.5	-	-
	300	-	-	44.7	33.4	-	-	65.4	-	-

Table 5. Linear algebraic solvers.

		DEC 8700	CONVEX C1 XP	CONVEX C210	FPS 64/60	ETA 10 MOD P	ALLIANT FX/4	IBM 3090/150	CYBER 990	CRAY XMP 14SE
CROUT METHOD	50	0.9	2.0	9.4	7.2	15.1	7.9	6.2	5.0	8.5
	300	0.8	3.7	26.9	11.0	52.1	7.6	12.0	9.8	26.5
GAUSS BLAS	50	0.9	2.0	-	5.4	5.0	1.0	8.4	4.9	16.7
	300	0.7	3.6	-	7.1	10.2	3.5	16.0	10.3	42.1
CHOLESKY BLAS	50	0.9	1.9	8.7	5.6	12.7	1.1	7.6	7.3	18.0
	300	1.0	5.0	17.3	6.1	61.8	4.5	14.5	23.2	58.3
USING SPECIFIC LIBRARIES	50	-	-	9.0	18.3	-	2.4	16.8	-	-
	300	-	-	34.3	31.8	-	16.7	38.0	-	-

Table 6. Application programs.

	DEC 8700	CONVEX C1 XP	CONVEX C210	FPS 64/60	ETA 10 MOD P	ALLIANT FX/4	IBM 3090/150	CYBER 990	CRAY XMP 14SE
HEAT EQUATION	77.7	148.5	-	54.2	29.5	-	33.9	43.	18.
		45.	14.8	32.0	69.6	30.3	9.9	-	-
FLUID DYNAMICS	2475.	3656	-	790.	-	3385.	514.	776.	143.
		408.	147.2	315.	69.6	381.	262.	446.	130.
POTENT. EQUAT.	32.6	24.8	6.9	9.7	20.7	23.0	7.1	18.0	4.7
OPERAT. RES.	118.0	209.4	78.7	570.0	-	375.9	63.1	55.0	30.
LIN. MEAN SQ.	318.7	347.5	74.9	79.1	125.5	476.5	68.6	52.0	18.7
REGR. ANALYSIS	18.0	9.0	5.6	2.5	2.3	-	-	3.6	3.0
CHEM. KINETICS	14.0	-	-	5.1	7.4	-	-	3.5	2.3

Remark: when there are two figures in one cell, the second refers to tuned versions of the code.

Table 7. Performance ratios, the DEC8700 being taken as a unit.

	DEC 8700	CONVEX C1 XP	CONVEX C210	FPS 64/60	ETA 10 MOD P	ALLIANT FX/4	IBM 3090/150	CYBER 990	CRAY XMP 14SE
BASIC OPERAT.	1	3.0	11.1	6.0	36.9	4.1	13.9	15.0	56.4
LIN. ALGEBRA	1	5.4	27.6	9.7	59.4	5.1	16.0	20.6	52.6
APPLICATIONS (*)	1 (**)	1.1	-	3.6	-	0.9	4.6	3.9	10.0
		2.2	6.9	4.6	8.7	2.8	6.6		
AVERAGES	1	3.2	-	6.4	-	3.4	11.5	13.2	39.7
		3.5	15.2	6.8	3.5	4.0	12.2		
WEIGHTED RATIOS	1	1.0	-	3.1	-	0.8	4.1	3.6	9.1
		1.9	6.3	4.0	7.6	1.8	5.8		

(*) without the "spaghetti operations" research program.
(**) second figures refers to tuned versions of the codes.

SOLVING PARTIAL DIFFERENTIAL EQUATIONS ON A NETWORK OF TRANSPUTERS

E. Verhulst

Intelligent Systems International, Zavelstraat 142

B-3200 Kessel-LO, Belgium

ABSTRACT

This paper describes in a first part some numerical experiments using the transputer. A simple equation (two-dimensional Laplace) was chosen as the vehicle for our efforts. This equation was solved using parallelized versions of the classical algorithms: Jacobi, SOR. Two different parallel methods were used for each of the two algorithms, allowing us to draw some more general conclusions regarding such performance-determining parameters as the computation/communication ratio in transputer networks.

We also speculate on the influence of a much faster floating-point unit on the performance of a network for this kind of problems (such a unit is now available in the form of the T800 Transputer).

In a second part an overview is given of an ESPRIT proposal that aims to design and implement a high level user-friendly programming environment for solving PDE on a network of transputers.

I. SOLVING THE EQUATION OF LAPLACE

Overview

We will start by describing some of the main characteristics of the INMOS Transputer, as far as they are necessary for a correct interpretation of the results and the performance obtained.

Then we discuss some of the differences between Occam, the language we chose to develop the software in, and more traditional languages such as Fortran.

Thirdly, we present some of the characteristics of the Transputer network we have been using.

In a fourth section we derive the linear equations system we used to obtain an approximation of the solution of the original Laplace equation.

In a fifth section we describe our two methods for parallellization of the solution process. The results we obtained and the observed performance of the algorithms are then discussed, and finally we state some possible future extensions and improvements of these implementations.

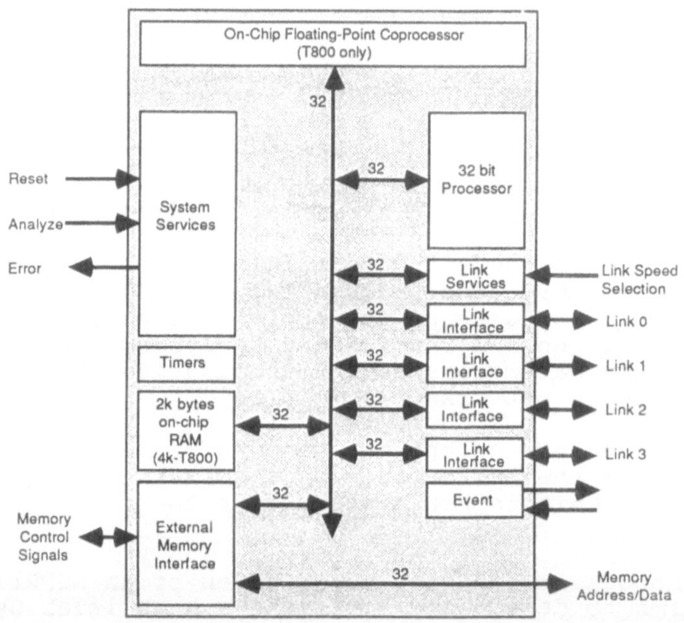

Inmos Transputer Block Diagram

Fig. 1.

1. The transputer

The transputer is a 32-bit microprocessor developed by INMOS Ltd. in the U.K. It is intended to be used in multiprocessor systems with local memory. A special characteristic in this respect is the presence of four serial links for interprocessor communication of the chip (see Figure 1). The maximum speed of these links is 20 Mbit/second. Another interesting feature is the presence, on chip, of 2 Kbytes of single-cycle memory. This means that the Transputer by itself is a complete, single chip computer. The peak performance of the 20 MHz version reaches 10 MIPS. An interesting recent development in the Transputer family is the incorporation of a very fast floating point unit on the chip (in the T800). This should provide for a performance of 1.2 Mflops or 4 Whetstones, which exceeds the performance of most processor coprocessor combinations. For the addressing of external memory, the Transputer uses a multiplexed 32-bit data-address bus. It should be noted that the internal memory of the Transputer is not a cache. Time slicing and task switching are performed by the hardware. Any process, and in particular communication processes, can be assigned a higher priority. Due to the concurrent handling of communication by the link-hardware, even at maximum communication rate, the CPU still achieves a 90% availability.

2. OCCAM

Occam differs from other languages mainly by the presence of constructs for the expression of concurrency in the language itself. The language does not always distinguish between concurrency on the processor and concurrency in multiprocessor environments. This means that communication "channels" are mapped on links when the communicating processes run on different processors, and are simply mapped onto memory locations when the processes run on one processor. The result is the logically constant behaviour of the user's program. The well chosen use of more than one Transputer assures improved performance. In implementation forms of Occam processes, this means that the PAR construct, which indicates that the processes may run concurrently, is implemented by time-slicing in the single processor case whereas in the multiprocessor case, the concurrency is real. Ultimately, this means that a program can be developed and tested on one processor, and when found to work correctly, can be ported onto a multiprocessor network without modifications of the program logic. However, the mapping of the different processes onto suitable processors and the mapping of the logical "channels" onto the physical links must still be done by the developer. At first sight, this seems a disadvantage. In software developing terms however, the programmer is freed from the sequential bottleneck and is able to concentrate on the conceptual development of his program. Very often, the result is modular and well structured software this much easier to understand and to maintain. In most other respects, Occam can best be compared to Fortran. Indeed, one might say that Occam aims to become the Fortran of the parallel world. For instance, all memory allocation is done at compile time. Bound up with this is the absence of recursion and the absence of the pointer concept. These constructs are planned to be introduced in future releases.

3. Hardware configurations

We had available one INMOS B004 board with one 15 MHz Transputer, having 2 Mbyte of external 7-cycle memory and one INMOS B003 board with four 15 MHz Transputers, each having 256 Kbytes of 4-cycle memory. The 4 Transputers on the B003 were connected in a simple chain. A PC served as host for the development system. All development was done using the TDS envi ronment of INMOS. It may appear that the applicability of the results is limited by the specific hardware configuration. This is, however, not the case. Firstly, the nature of the topology is such that extension is a trivial matter. Secondly, the influence of the slow memory is fairly simple to quantify, due to the absence of such "features" as cache memory, memory management units and the like. However, it would have been interesting to be able to use the floating-point version of the Transputer. Figure 2 represents the hardware configurations used.

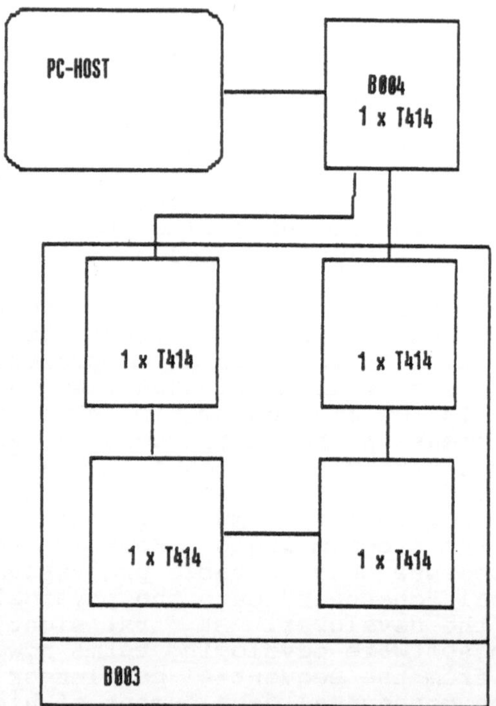

Fig. 2. Hardware configurations.

4. Derivation of numerical approximations

The following is our target equation:

$$\frac{\partial^2 w}{\partial x^2} + \frac{\partial^2 w}{\partial y^2} = 0$$

Start by approximating the derivative $\frac{dw}{dx}$ by the Taylor series expansion, where w is the function value at a point x near x_0:

$$w = w_0 + \frac{dw}{dx}(x-x_0) + (\frac{1}{2}!)\frac{dw^2}{dx^2}(x-x_0)(x-x_0) + \ldots \quad (1)$$

By substituting $w_1 = x + h$ and $w_3 = x - h$ into this, we obtain:

$$w_1 = w_0 + h\frac{dw}{dx} + h.h/2\frac{d^2w}{dx^2} + \ldots \quad (2)$$

and

$$w_3 = w_0 - h\frac{dw}{dx} + h.h/2\frac{d^2w}{dx^2} + \ldots \quad (3)$$

Adding (2) and (3) we get:

$$w_1 + w_3 = 2.w_0 + h.h\frac{d^2w}{dx^2} + 0(h^4) \quad (4)$$

where $0(h^4)$ is a term that includes fourth, sixth and higher order terms. For a well chosen h we can neglect this term.

Rearranging the equation (4) yields:

$$h.h\frac{d^2w}{dx^2} = w_1 + w_3 + 2w_0 \quad (5)$$

For two dimensions, we get:

$$h.h\frac{\partial^2w}{\partial x^2} = w_1 + w_3 - 2w_0 \quad (6)$$

$$h.h\frac{\partial^2w}{\partial y^2} = w_2 + w_4 - 2w_0 \quad (7)$$

Multiplying the original equation by h.h and substituting these two difference equations into it gives:

$$w_0 = \frac{1}{4}(w_1 + w_2 + w_3 + w_4) \quad (8)$$

or in words: the value of any gridpoint can be linearly approximated by the arithmetical mean of its four rectangular neighbours. The simple Jacobi algorithm calculates this value iteratively for each point. The SOR (Successive Overrelation) algorithm calculates alternately the odd and even set of points. For details see Ref. [1]. The fact that this linear approximation is simple to implement, makes the equation of Laplace an ideal laboratory vehicle for our purpose:

- the computational effort and the algorithms are simple;
- the computation to communication ratio is critical and directly influences the global performance.

5. The parallel algorithms

5.1. The packet-sending method. This was the first method we implemented. It was based on the idea of "dynamic load balancing", that is to say: we try to keep every processor as busy as possible at runtime. This idea appears to be useful mainly in cases where runtimes for a given "packet" of calculations vary unpredictably (a popular example of this is the calculation of Mandelbrot sets: the number of iterations for each point in the complex plane is not known beforehand). In our implementation of this idea, the B004 Transputer acts as a master while all B003 Transputers (configured in a pipeline) run identical programs. On a B003 processor, a calculator process is running in parallel with a routing process. The calculator accepts "packets" of data (prepared by the master B004) consisting of the four neighbouring values of a grid point. It calculates the arithmetical mean of these numbers and passes the result (which is the next value of the grid point) to the following processor in the chain. The routing process, on the other hand, directs incoming packets either to the local calculator process (if it was not busy) or else (this means, if the calculator was already busy or if the packet had already been treated) it passes it on to the next processor via a multiplexing process.

The B004 for its part prepares packets, sends them to the network, and accepts the results on the other side of the pipeline. The full grid is recalculated on each iteration until the converge criterion is met. A variation on this idea is to let the packets contain an identifier which "bounds" them to execution on a specific processor in the pipeline. The computer then keeps the packets for execution on its processor, or sends them through, without regard for busy or free processors. To provide an example of Occam code, the program code is given below:

```
PROC node.process(VAL INT node.number,
                  CHAN OF package in, out)

   ... PROC demultiplexer (CHAN OF package in, out1,out2)
          -- code has been folded away

   PROC calculator (CHAN OF package in, out)
     INT i :
     [5]REAL32 string :
     SEQ
       i := 0
       WHILE i <> stop.signal
         SEQ
           in ? i; string
           IF
             i <> stop.signal
               SEQ
                 string [0] := (((string[1] + string[2]) +
               string [3]) + string [4])/4.0(REAL32)
                 out ! i; string
             TRUE
               out ! i; string
   :
   ... PROC multiplexer (CHAN OF package in1,in2, out)
          -- code has been folded away

   CHAN OF package to.calculator :
   [2]CHAN OF package to.multiplexer :

   PAR
     demultiplexer(in,to.multiplexer[0],to.calculator)
     multiplexer(to.multiplexer[0],to.multiplexer[1],out)
     calculator(to.calculator,to.multiplexer[1])
:
```

As was to be expected in view of the limited amount of computation that had to be performed on each packet, saturation in the pipeline was very quickly reached. Two main factors contribute to this:

1. The B004 master Transputer has to receive and prepare all the packets, resulting in a bottleneck. In our case, with the SOR algorithms requiring some more instructions than the Jacobi algorithm, only two of the four Transputers of the B003 pipeline could be kept busy.
2. This effect was even more pronounced due to the inherently slower operation of the B004 Transputer. On this Transputer, most data had to be read from and written to the slow off chip memory, while on the B003, all code could be kept on the fast on chip memory.
3. In general, this approach leads to a serious overhead in communication. In effect, at every iteration step, each packet has to pass through the whole pipeline. Note that the effect would have been even more pronounced in the presence of a hardware floating point unit.

To conclude, calculator processes were sitting idle, waiting for data to arrive through the pipeline. So a logical step was to reduce the amount of communication. This is the basic idea of our next algorithm.

5.2. Distributed calculation method. In this method, the role of the B004 processor remains comparable to the previous algorithm. However, instead of constructing a packet per grid point, the B004 now splits the total grid into four subgrids and downloads the four subgrids to the four B003 processors. This means that communication is much less than in the first algorithm. Indeed, in the first method each gridpoint appeared in four packets, whereas here it only appears in one. Additionally, since there are far less communications in this case, we improve the global computation to communication ratio. Of course, we now have to add the exchange of subgrid boundaries and a partial convergence sum between the B003 processors to our total number of communications. But since this amount increases only linearly with the size of the grid, while the previous communications increase quadratically, this is only a second order effect. So we conclude that much less time is spent communicating than in the previous example. This is certainly a desirable result, but have we increased the computation/communication ratio? Indeed we have, for the total number of calculations has not changed from the first algorithm, except for some linearly increasing additional calculations involving subgrid boundaries. The observed results confirmed our expectations: except at the startup no processes were idle for this implementation.

6. Results and discussion (see Table 1)

The best results were obtained by using the red-black SOR algorithm in combination with the distributed calculation method. A grid of 22x78 points took 5.47 seconds to calculate on the Transputer network, compared with 250 seconds for the sequential version on a 7 MHz PC/AT. Compare this with the results obtained by the Jacobi algorithm, which took 573 seconds on the network. As is already known, it is still better to have a fast algorithm than to have more processing power.

Table 1. Some numerical results (ε = .00001, 22x78 grid-points, 4 byte reals)

Algorithm used:

1. Jacobi
   1 T414.15 : Package method: 32 m 44 s, 2365 iterations
   4+1 T414.15 : Package method: 11 m, 2365 iterations

2. Red-black SOR (WF = 1.65)
   1 T414.15 : Package method: 1 m 18 s, 87 iterations
   Distributed method: 38 s, 87 iterations
   4+1 T414.15 : Package method: 18 s, 87 iterations
   Distributed method: 6 s, 87 iterations

3. Red-black SOR (WF = 1.65, with saving partial sums)
   1 T414.15 : Distributed method: 35 s, 87 iterations
   4+1 T414.15 : Distributed method: 5.47 s, 87 iterations
   PC/AT/7 MHz : 5 m 20 s, 112 iterations (6 byte reals)

4. Variant of SLOR (WF = 1.87) (distributed calculation)
   1 T414.15 : 29 s, 65 iterations
   4+1 T414.15 : 5.29 s, 65 iterations
   PC/AT/7 MHz : 4 min 20 s, 89 iterations

The results favour the distributed calculation method in this case, but can we draw any more general conclusions from this? Certainly it is so that the packet sending method is simpler to implement and also far easier to extend for use on larger numbers of Transputers. Because all Transputers have an identical role, the processes running on them do not have to know anything about the geometry of the problem; whereas in the distributed calculation method, communication of subgrid boundaries is different from processor to processor depending on the particular subgrid it is calculating. Also the problem arises of mapping subgrids to particular processors depending on the communication network, which is irrelevant in the packet sending method. However, in the latter method, for this particular amount of computation (computing the arithmetical mean of four numbers), no further speedup is reached when more than 2 Transputer on the D003 board are used. This certainly does not mean that we have to reject the packet sending idea for other types of problems: it has, after all, the very desirable characteristics of topology independence and processor equivalence. One could certainly use this idea to good effect for certain applications in image processing, where communications between adjacent parts of the image are not required: in that case it is very simple to increment the amount of computation per packet above some critical threshold by increasing the amount of image pixels per packet.

As a general conclusion, we might state that it is important to limit the number of interprocessor communications to a minimum, or stated differently, that it is a good idea to avoid using non-local data as much as possible. A conclusion that is very specific to the Transputer concerns the use of the chip's internal memory. It would certainly be a good idea to have all frequently used data in the on-chip memory; however, since these data are not necessarily identifiable at compile time and since there might be more than can be statically contained in the internal memory, it it very difficult to use this feature of the Transputer. This is unfortunate, because drastic performance improvements would certainly result from using this memory: it is at least three times faster than off-chip.

Note also the results obtained with a variant of the SLOR algorithm with a weight factor. Although the number of iterations needed is significantly less, the total computation time is about equal to the red-black SOR algorithm. At first sight, this seems surprising. Further investigation however revealed that this was due to the fact that the programming of the SLOR algorithm required more control operations and thus more "overhead" for each iteration. This indicates how sensitive the performance is to small variations of the code and that algorithms that have a better convergence rate but that are more complex to program, can result in a lesser overall performance. We remind that this conclusion does not take into account the number of gridpoints involved as this was kept constant.

II. A HIGH LEVEL PROGRAMMING ENVIRONMENT FOR PDE

Overview

A demonstration procedure program with a user-friendly interface has been developed for solving the equation of Laplace in two dimensions. This approach will be extended towards the design and implementation of a generic programming environment for solving PDE in three dimensions on a parallel computing network. It will provide the application developer with an open environment for programming dedicated applications modelled by PDE. End users will use the generated instance of the environment as an user-friendly stand-alone application. This project is the subject an an ESPRIT II proposal.

1. Rationale for the project

Current users of supercomputers are faced with the dilemma of continuing to use their existing software (very often written in Fortran) on the available but expensive supercomputers (such as CRAY, CYBER 205 or CONVEX, et al.) or to rewrite their application programs for a more affordable parallel computer. Although current parallel computing networks provide a cost performance ratio which is between 10 to 100 better, the step is still not easily taken as the step is likely to take several manyears. Moreover, a lot of supercomputer users do not write the software themselves, but use packages developed by third parties. At the same time, it seems almost contradictory that while computing power is becoming available at a few 100 $/Mip, the programmer is still forced to struggle with the same programming tools as before. The conclusion is that in order for parallel computers to become successful, the user should experience a similar level of improvement on the software side. This is less trivial to do, especially in the case of parallel computing networks. The main reason is that this calls for conceptual breakthroughs. One solution is the development of implicit parallel declarative languages. Although this can be done, the implementations are still not competitive enough in terms of performance compared with conventional imperative languages. For this much higher communication bandwidth between cooperating processor nodes in a network are needed. Current state of the art indicates that these developments are about 5 years off a practical usability. In the case of PDE however, the degree of parallelism is often highly regular and very often is the topology of the physical problem to be solved a direct reflection of the geometry of the problem. Thus, the mapping of the problem topology into the computing network is a static one with a regular and predefined communication pattern between the processing nodes. As a result, the conditions are met so that higher level programming tools can be developed with the aid of imperative (and conventional) languages.

The user is able to concentrate on the actual algorithm of his problem as all the codes for defining the geometry, the graphical representation and the details of the mapping are predefined. The result is that within the particular problem domain, the parallelism involved is transparent whereas the development cycle is drastically reduced.

As an analogy, we can consider spreadsheet and database programs in the PC market. They enable managers with no computer science background to "write" programs (in a particular domain) instead of weeks using high level languages such as Pascal. This kind of open ended business applications programs are largely responsible for the success of personal computers.

2. Architecture of the environment

The software architecture is composed of a number of layered modules. The three main parts are: the user interface, the code generator kernel and the network manager. Although the current projects targets Transputer networks, the portability to other parallel computing networks (based on message passing) is a major objective of the project.

2.1. The user interface. Whereas the computing performance is determined by the implemented algorithms and the amount of available MFlops, the user interface greatly influences the design or simulation productivity of the user of the system. Therefore, this is an important part of the system. It will consist of a number of overlapping windows as can be found on most of the advanced workstations. The geometry will be defined by a "standard" CAD system that can be called in one of the windows. The resulting CAD file needs to be exported into the PDE-datastructure, which is basically a representation of the three-dimensional grid. The user must be able to change locally the resolution of the grid and can define the boundary conditions. The system should provide a number of predefined menus so that the user can select the appropriate problem domain and the desired algorithms. Application developers must be able to add new applications and algorithms. Another important part of the user interface is the graphical representation window. This must enable the user to rotate the view, shift the viewpoint and to zoom in on any part. Individual gridpoints can be inspected in detail.

2.2. The codegenerator kernel. This module is the kernel of the system and must combine the PDE data structure with the chosen algorithm into bootable code. The idea is that PDE very often result in the iterative solution of a set of linear equations on each gridpoint that is dependent on the problem domain and the chosen algorithm. On the other hand the code that handles the interprocessor communication, the input/output, is more static and can be written in a generic way which is dependent on the problem and network topology. Therefore, the user should not be concerned with this code as this can be precoded by the application developer. In our case we will use standard language utilities that are invoked in the background.

2.3. The network manager. Although its function is equivalent to the function of a normal operating system, the user of the latter term leads to confusion. In a parallel computing network, the task of the operating system is to ensure that the communication between the processor nodes is carried out and most importantly that the "CPU" resources are attributed in accordance with the user's request. Additional functions include error reporting, handling the input/output with the filing system and configuring the linkswitches so that the desired network topology is obtained. In our case, using a general purpose operating system (such as the distributed UNIX-like HELIOS) that enables transparent internode communication and dynamic p processor application, would result in a serious overhead as most of these features are to be carried out in software on the current version of the Transputer. As a large network needs to be shared among several users, this will be implemented by dividing the network in different user domains. Thus the network manager will provide for a rather static handling of the computing resources. This is largely counterbalanced by the improved performance and is justified by the probably dedicated use of the computing network.

3. Additional design of objectives

One has to be aware that although the programming environment, or at least a subset of it, can be realized with the existing hard- and software technology, the final performance is still influenced by a number of system dependent features. These are shortly discussed below.

3.1. Input/output. The targeted applications will in some applications call for several 100 Mbytes of data to be exported and imported between the user and the network. In the current systems, this amounts to a bottleneck providing a few 100 Kbytes/sec. (A new VAX-interface card reaches a peak of 1 Mbytes/sec.). Thus solutions have to be found for diminishing the bottleneck. The same problem is encountered when some algorithms call for global communication between the iterative steps. As a conclusion, one can say that it would be desirable to implement a separate communication "highway" for input/output and global communication besides the normally available links for local communication. Solutions can be designed, but they still pose a challenge in terms of a minimum of added cost and complexity to the resulting system.

3.2. Error recovery and reporting. Currently, the transputer has limited means for the detection of runtime error and reporting. No hardware provisions have been made for error recovery. Thus, when a runtime error occurs, the system is halted due to a resulting deadlock situation. The only means of cure are a design of the algorithms that detects possible runtime error in complement of the compile time checks, when nevertheless an error occurs (raising) the error signal. This must be monitored by the network manger. When this is reported to the host, it is possible to invoke the network debugger so that the failing source statement is detected. In the case of application end users, this will result in calling the service center of the organization that developed the application.

REFERENCES

1. R.W. Hockney and C.R. Jesshope, "Parallel Computers", Adam Hilger (1981).
2. The Transputer Reference Manual, INMOS Documentation.
3. Occam 2 Reference Manual, INMOS Ltd., Prentice Hall (1988).
4. M. Homewood, The IMS T800 Transputer, IEEE Micro, October 1987.
5. D. Reed and R. Fujimoto, Multicomputer networks. Message based parallel processing, M.I.T. (1987).

IV. INDEXES

AUTHOR INDEX

SUBJECT INDEX